Water Management in the 21st Century

NEW HORIZONS IN ENVIRONMENTAL ECONOMICS

General Editors: Wallace E. Oates, *Professor of Economics, University of Maryland, USA* and Henk Folmer, *Professor of Economics, Wageningen Agricultural University, The Netherlands and Professor of Environmental Economics, Tilburg University, The Netherlands*

This important series is designed to make a significant contribution to the development of the principles and practices of environmental economics. It includes both theoretical and empirical work. International in scope, it addresses issues of current and future concern in both East and West and in developed and developing countries.

The main purpose of the series is to create a forum for the publication of high quality work and to show how economic analysis can make a contribution to understanding and resolving the environmental problems confronting the world in the twenty-first century.

Recent titles in the series include:

Making the Environment Count
Selected Essays of Alan Randall
Alan Randall

Controlling Air Pollution in China
Risk Valuation and the Definition of Environmental Policy
Therese Feng

Sustainable Agriculture in Brazil
Economic Development and Deforestation
Jill L. Caviglia

The Political Economy of Environmental Taxes
Nicolas Wallart

Trade and the Environment
Selected Essays of Alistair M. Ulph
Alistair M. Ulph

Water Management in the 21st Century
The Allocation Imperative
Terence Richard Lee

Institutions, Transaction Costs and Environmental Policy
Institutional Reform for Water Resources
Ray Challen

Valuing Nature with Travel Cost Models
A Manual
Frank Ward and Diana Beal

The Political Economy of Environmental Protectionism
Achim Körber

Water Management in the 21st Century: The Allocation Imperative

Terence Richard Lee

NEW HORIZONS IN ENVIRONMENTAL ECONOMICS SERIES

Edward Elgar
Cheltenham, UK • Northampton, MA, USA

Published by
Edward Elgar Publishing Limited
Glensanda House
Montpellier Parade
Cheltenham
Glos GL50 1UA
UK

Edward Elgar Publishing, Inc.
136 West Street
Suite 202
Northampton
Massachusetts 01060
USA

A catalogue record for this book
is available from the British Library

Library of Congress Cataloguing in Publication Data

Lee, Terence R.
  Water management in the 21st century: the allocation imperative /
  Terence Richard Lee.
    (New horizons in environmental economics)
  1. Water-supply—Economic aspects. 2. Water Utilities—Deregulation.
  3. Privatization. 4. Water-supply—Economic aspects—Developing countries.
  5. Water utilities—Deregulation—Developing countries. 6. Privatization—
  Developing countries. 7. 21st century—Forecasts. I. Title. II. Series.

      HD1691.L44   1999
      333.91'009172'4—dc21   99–042179

ISBN 1 84064 080 4

Printed and bound in Great Britain by Biddles Ltd, www.Biddles.co.uk

Contents

List of Tables

List of Figures

Introduction

Who would not choose to follow the sound of running waters? Its attraction for the normal man is of a natural sympathetic sort. For man is water's child, nine-tenths of our body consists of it, and at a certain stage the foetus possesses gills. For my part I freely admit that the sight of water in whatever form or shape is my most lively and immediate kind of natural enjoyment; yes, I would even say that only in contemplation of it do I achieve true self-forgetfulness and feel my own limited individuality merge into the universal.

Thomas Mann, *A Man and His Dog*[1]

We all tend to have romantic ideas about water whether for the reasons Thomas Mann suggests or for others. There is a history in all societies, at least, in all the societies that I have any knowledge of, of perceiving water as a gift from the gods. If traditionally society has felt, therefore, that water should be a free good, it is not surprising that these feelings linger. Nor is it surprising that even today we find it difficult to treat water as a commodity, as an economic good, even when we recognise that it is a scarce resource. On occasion, in fact, quite the reverse has been the case: the existence or perception of scarcity has been used as an argument for not treating water as an economic good. There is a strong emotional feeling that water is just too precious and essential for human survival to be treated as merely a commodity. Water is, however, a very practical and necessary resource and there can be little doubt that within modern economies, at whatever level of development, water is a commodity and it is almost always for economic purposes that we use water. Of course, at the same time that it is an economic commodity, water is also a splendid part of the natural environment and of some of the world's most attractive landscapes, but then

---

[1] From the Vintage Books Edition, 1954, translated from the German by H.T. Lowe-Porter.

so are many other resources, and these, including land, most societies do treat as economic goods.

In its character as a natural resource, water is scarce as are, by definition in economic theory, all resources used to satisfy human wants. We use water, as we use all resources, to satisfy our wants and because our wants are, more or less, unlimited then we have a problem of choice, of choosing how to employ the resources to satisfy which demands among the large number of individual and social wants.

This is a book, therefore, about choices, specifically, a book about choices in water management. It is about how to make choices and how to establish an institutional environment so that the choices are made in a rational and transparent manner. In order, however, to place the discussion about making choices in water management in context, the book also must at least touch on the nature of water as a resource, as well as on how man is using it. The bulk of the discussion, however, deals with the question of how we can use water more efficiently and equably. The issue of the efficient and equable allocation of water can be expected to be the dominant theme in water management in the next century. The bias in the discussion presented in the book is clearly towards moving the decision-making process on water allocation into the market-place and, conversely, taking it as much as is possible out of the political arena. The decision to leave water allocation to the market and of, therefore, treating water as an economic commodity also implies increasing the participation of the private sector in water management. This in turn means changing the role that governments have traditionally tended to play in water management.

In this book, I do not propose to test a hypothesis with any scientific rigour. However, underlying the discussion there is the implicit hypothesis that water is not scarce, in the sense of absolute water scarcity, but rather that we face an unavoidable challenge to radically improve the management of the allocation of water among uses. Chapter 1 deals explicitly with the question of water scarcity beginning with a discussion of the nature of water as a resource and continuing with the consideration of the reality of the commonly discussed issue of water scarcity. In this discussion, questions are raised such as, is water now considered scarce because we are running out of it, or is water now considered scarce because we are now realising for the first time that it is necessary to treat it as an economic good?

Last year a column in *The Times* newspaper had as a subheading the phrase 'If desalination were cheaper, we could drink 99% of the world's water' (*The Times*, 1998). What the article should have had as the subheading is, however, 'When desalination **is** cheaper, we **will** be able to drink 99% of the world's water'[2]. So, as will be explored further in Chapter 1, we cannot be said to be running out of water, but the cost of obtaining fresh water, at the time we want and in the quantities we want, is rising in many places around the globe. We are beginning to have to make serious choices about distributing water among uses and between the resources we devote to using water and to the use of other commodities.

Chapter 2 examines, in a brief historic survey, the nature of the institutions which different societies have developed to manage water. Most of the discussion is devoted to recent experience, over the last half-century or so, although reference is also made to older civilisations. Over the last 50 years, the choice has been, as in most of the history of water management, to use a system of centralised, bureaucratically controlled institutions. This choice is contrasted with an alternative decentralised, or local, system based on institutions where the users are dominant. This alternative, which has always been available and used, is now becoming, at the turn of the century, the preferred system for water management in many countries.

The remaining three chapters in turn discuss the main factors involved and to be considered in moving the process of the allocation of water to the market and of increasing the extent of private participation in water management. Chapter 3 analyses the ways and means for establishing and operating effective water markets and describes the experience of those few places where water marketing has been used as the method of water allocation. Chapter 4 explores, in some detail, the alternatives for increasing private participation in water management, in particular for the incorporation of the private sector in the ownership and administration of water-related public utilities. Finally, Chapter 5 discusses the repercussions of greater private participation on the public sector. The necessary elements to be including in policies for privatising public sector monopolies are presented together with the means for establishing and running effective regulatory systems, once these monopolies have been transferred to private ownership.

This book concentrates almost entirely on the issue of how to better and more efficiently allocate water through changing water management institutions. There is little discussion of other aspects of water management, such as the not insignificant question of how to manage water quality. This is a deliberate omission. An omission made not because water quality is not important, but because in the opinion of the author the management of water quality is not yet so susceptible to improvement through the introduction of market processes. The basis of this belief is the fact that improvements in

---

[2] Matthew Parris, "Every drop to drink", *The Times*, Internet Edition, 14 August, 1998.

water quality in most countries, except for the most highly developed, require previous considerable investment in sewerage and waste treatment plants, particularly to treat domestic wastes. Water quality in most countries would be considerably improved through treatment of domestic sewage. Once the responsibility for sewerage systems is in private hands or in the hands of efficiently run public companies and a proper tariff charged, then they will be built. Today, the major contributor to poor water quality in most developing countries is the inadequate management of sewerage by the public sector and the lack of investment in waste treatment both for domestic and industrial wastes. It is the inadequate deposition of sewerage from sewer systems and from badly maintained latrines in urban areas that is the major cause of water pollution. Industries, too, contribute, but efforts to convince industry to reduce its contribution, either through changes in processes or through treatment, tend to be wrecked on the rock of the public sector's failure to effectively manage domestic sewerage.

In this context, it seems inappropriate to discuss the use of economic or market instruments for the management of water quality. The first need is to create the infrastructure that will make the use of economic instruments both possible and practicable. This is why the important and growing challenge posed by the deterioration of water quality in nearly all countries is not explicitly addressed in this book. It will be left to a possible future work so that there is no confusion about the real current advantages of applying economic instruments to the management of water allocation.

This situation in respect of water quality contrasts markedly with the issues posed by the ineffective management for the allocation of water and for the administration of water-related infrastructure and services. Both primordial tasks of water management are more than ready for the application of economic instruments and the transfer of their administration to the private sector.

Much of the material for this book has been based on studies made while I was a staff member of the United Nations Commission for Latin America and the Caribbean (ECLAC). The experience I gained while working there for more than 25 years is the real foundation on which this work has been built. Much of the material has benefited significantly from the innumerable discussions with my colleagues at ECLAC. I would like to acknowledge the assistance I received over the last 10 years of my career in the United Nations from Andrei Jouravlev who is an indefatigable researcher and assembler of bibliographical references. I must also give thanks to Professor Henk Folmer who first suggested that I contact Elgar Publishing about the possibility of producing this book.

Finally, my work has benefited enormously from the continual discussion of the ideas expressed with my wife, Frances, and from the corrections made by her and my stepdaughter, Mary, to my often careless prose style.

1. Confronting the Scarcity of Water

Water, water, everywhere
And all the boards did shrink;
Water, water, every where,
Nor any drop to drink.

Samuel Taylor Coleridge, *The Rime of the Ancient Mariner*

There is a widespread and growing opinion that the world faces an imminent and ever more serious water crisis. This opinion is shared by many of the professionals involved in water management as well as much of the public at large, because these opinions are widely expressed. This crisis is presented not just in the sense that water as a resource will become a scarcer commodity than it has been historically, but in the sense that the scarcity will be so marked as to cause serious economic, political and social repercussions. There is an increasing number of papers being presented to academic conferences, and there are articles in the press and official reports which suggest that the threat of increasing water scarcity is becoming so grave as even, as stated in a relatively balanced recent report from the United Nations, to be 'weakening one of the resources bases on which human society has been built' (WMO, 1997). Other recent reports are less cautious and speak of increases in water scarcity leading to 'conflicts among users, regions and countries' (IPCC, 1997).

In the absolute sense, however, even given climate change, water is not becoming scarcer. It is true that the amount of water is finite, but it is not only finite, it is inherently stable. We are not using up the water resource, because we cannot use it up. Water is not exhaustible in the sense that coal, oil, any metal ore or other non-renewable natural resources are. The amount of water available on earth has not changed for a very long time. Why then is there this concern about water scarcity?

The present concern with growing water scarcity seems to be due to the interaction of a multiplicity of very different factors. Among these factors the most important seem to be the following:

1. the concept of water as a finite and limited resource;
2. a tradition of a bureaucratic and centralised approach to water management;
3. a concern with the continuing growth of the human population;
4. the increasingly widespread economic change and growth in a single global economy that is reflected in the expansion of industrial, urban and irrigation demands for water.

There is, in addition, a strong tendency by many commentators to accept a Malthusian view of the relationship between water and population which has been strengthened by the uncertainties introduced by the possibility of climate change due to global warming.

In this chapter, each of these issues will be discussed in turn to reach some conclusion as to the reality of the issue of water scarcity. This is a necessary basis for the presentation of management alternatives to allocate water efficiently which is the core of the discussion in this book.

THE QUANTITY OF WATER ON EARTH

It is a tradition that books on water management always begin with a discussion of the quantity of water and its distribution. Then the authors continue by pointing out that nearly four-fifths of the surface of the earth is covered by water, by the oceans. The discussion generally proceeds to draw the reader's attention to the contradiction that only a small proportion of this water, 2.5 to 3 per cent, is freshwater, and, therefore, directly suitable for the maintenance of human life. For what has been important to man is not the existence of the water in the oceans, but the hydrological cycle which begins with the evaporation from the oceans of salt-free water, much of which falls on land in the form of rain or snow before flowing back, by one means or another, to the oceans to begin the process over again.

The circulation of freshwater through the hydrological cycle is a major element in the maintenance of the natural environment, as we know it. The

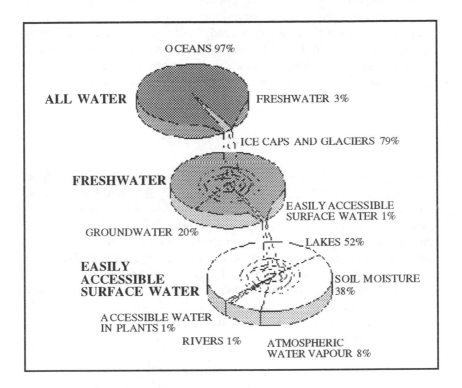

Source: Lean and Hinrichsen (1994)

Figure 1.1 The distribution of water

hydrological cycle is made up of a number of components and human society has historically made use of various of these components, in particular snow, runoff from precipitation, soil moisture, groundwater and temporary storage in lakes or aquifers (Figure 1.1). The total amount of freshwater accessible in from these sources for human use is, however, only a very small proportion of the total amount of freshwater that is found on earth. Indeed, it is equivalent to only about 1 per cent of that 2.5 per cent of the water on earth that is not salt. This is because the largest proportion of freshwater is trapped in the permanent ice and snow of the polar ice caps.

Evidence of exactly how much freshwater there is and of its distribution is not conclusive. The lack of exactitude is not, however, of serious significance. For whatever the amount and distribution is, and even accepting that only a small proportion of the total amount of water is readily accessible for human use, there is a still great deal of freshwater suitable and accessible for human use. It is estimated that the 1 per cent available is equal to some 500,000 cubic kilometres of water a year (Pereira, 1973).

THE AUTOCRACTIC INHERITANCE

It is, therefore, not surprising on the one hand, that given the large amount of water generally available it has in many parts of the world been taken for granted historically as a gift of the gods. Water has always been for most human societies the most accessible and plentiful natural resource. On the other hand, there have been significant and historically influential societies where water was scarce. These were very important exceptions in the history of human development, because it is generally accepted that modern civilisation has its origins in water-short societies. The need to manage water, to collect and to distribute it led to the development of the great hydraulic societies in Egypt, Mesopotamia, Northern China, Central Mexico and the coast of Peru.

> In a landscape characterised by full aridity permanent agriculture becomes possible only if and when co-ordinated human action transfers a plentiful and accessible water supply from its original location to a potentially fertile soil. When this is done, government-led hydraulic enterprise is identical with the creation of agricultural life. This first and crucial moment may therefore be designated as the 'administrative creation point.'
>
> Having access to sufficient arable land and irrigation water, the hydraulic pioneer society tends to establish statelike forms of public control. Now economic budgeting becomes one-sided and planning bold. New projects are undertaken on an increasingly large scale, and if necessary without concessions to the commoners. The men whom the government mobilised for corvée service may see no reason for a further expansion of the hydraulic

system; but the directing group, confident of further advantage, goes ahead nevertheless (Wittfogel, 1957).

These societies were based on centralised hydraulic systems. They were autocratic and bureaucratic, with a complete centralisation of authority and power, and developed very sophisticated and complex social structures that needed mathematics and writing to manage water. In all such societies, despite the variations in form, the water control, transport and distribution systems were run through systems of complex and detailed bureaucratic administration. This bureaucracy became all-powerful, commonly in the form of priestly castes as the educated class. This educated cadre was needed to keep track of the material and human resources at the disposal of the state for constructing and running the water control and distribution systems.

The Inca Empire, for example, was structured into five caste levels. At the top was the Inca and his four member council of state. The second level was made up by the imperial nobles who occupied the superior administrative posts. The third level consisted of the *curacas* or *caciques* who were responsible for running communal works and land distribution fundamental to the operation of the irrigation systems. The remaining two castes, known as the *ayllus* and their servants, formed the majority of the population and performed the work (Chonchol, 1994).

The core of Wittfogel's classic study was to explain the influence of the needs imposed by the dependence on irrigation systems on the development of despotic states. However, the reason for bringing this discussion in here is to point out how this ancient bureaucratic centralised approach to the management of water has had a strong indirect influence over modern water management. This influence occurred despite the decline and disappearance of the great hydraulic societies long before modern water management developed and the existence of alternative solutions.

The existence of an alternative approach towards water management permits the classification of modern irrigation societies into two general types on the basis of their historical origin. On the one hand are those societies which have grown piecemeal as individual and collective private initiative has built water control and distribution systems with little or no central direction from the government, as in much of Western Asia, Southern Europe, Taiwan, and parts of Latin America. On the other hand, there are those societies created as deliberate attempts to develop new economic opportunities based on irrigation. These modern attempts to develop hydraulic societies in water-scarce regions have tended to revive the relationship observed by Wittfogel between authoritarianism and water management. There are various examples: the development of large irrigation schemes in the Indus basin by the British authorities in the Punjab in colonial India; the major effort undertaken in Mexico to develop agriculture on the basis of irrigation after the revolution; the huge projects in the Central Asian republics of the former Soviet Union and; such influence can be seen even in the projects of the Board of Reclamation in the western United States. Beginning in the 1930s, and best exemplified by the Tennessee Valley Authority (TVA), the influence of the centralised, bureaucratic approach to water management grew into the new idea that the unified management and development of river basins could form the basis for the growth of poorer regions and countries. Such policies became almost universal once the multilateral development agencies adopted them in the 1950s and applied them to many projects in developing countries. Until recently, the belief that river basin development was a means to more general development came to be widely held in the United States in particular:

> Following the enactment of the TVA Act, proposals for the creation of similar government agencies followed thick and fast.... In the United States, proposals ranged from a general plan for subdividing the United States into valley authorities to individual agencies for the Columbia, the Missouri and the Arkansas (White, 1957).

White went on in this paper to comment that by 1957 the idea had been 'adopted sparsely' outside the United States.

This, however, was to change as the influence of the multilateral lending agencies made itself felt in the world. In a later article, White noted that 'the pace of public investment in water management has accelerated and United Nations involvement in promoting both national and international action in river development has expanded widely' (White, 1970). Perhaps this expansion would not have occurred, or the bureaucratic approach to water management not have remained so common, were it not for what can only be described as the quasi-religious view of water as a resource apart from other resources. This view justifies the need to treat water differently and, therefore, to manage water use bureaucratically. The arguments all feed one on the other, but the result is quite clear: water management cannot be left to the decision of the users and the market.

It must be recognised that treating other resources as economic goods does not stop the periodic appearance of fears of scarcity or of the rapid diffusion of ideas that we are about to run out of them. The most recent of such events occurred with the oil crisis in the 1970s, which led to the propagation of all types of dire predictions on the consequences of the world's dependence on oil. In the event, the result of the attempt by the OPEC nations to control the oil market ended in a market adjustment, both in supply and in demand, which undercut the very market power that they were attempting to exert by creating an artificial scarcity of oil. However, the idea of an oil crisis persists, as shown by a recent article in *Science* that argues that the next and largest, if not terminal, oil crisis is all but upon us (*Science*, 1998).

EXPANDING POPULATION AND WATER SCARCITY

The belief that there is an imbalance between the economic availability of natural resources and population growth has been widespread in modern society (Barnett, 1960). The obvious resource-saving nature of modern technology, in general, and the emergence of the computer, in particular, has over the last quarter of a century pushed these ideas into the background. In fact, the resolution of the oil crisis seemed by the beginning of the 1990s to have effectively driven any discussions of increasing resource scarcity acting as a brake on human development away from centre stage. Perhaps to an extent that had not occurred since the Reverend Thomas Malthus published his 'An Essay on the Principle of Population', there has been little attention paid recently to any limits on economic growth posed by resources. It is, however, precisely in the debate about the existence or not of a water crisis that Malthusian ideas are again resurfacing at the centre of the discussion.

The Influence of Malthusian Ideas

In this debate, the views on the nature of the challenge posed by the realisation that water is a scarce good can be divided into what can be termed as 'neo-Malthusian', on the one hand, and the view that can be termed 'neo-classical', on the other. Without entering a detailed discussion of economic theory, the differences between the two can be said to lie in the different perception of whether in the long-run marginal productivity increases. Those holding neo-Malthusian views, as did Malthus himself assume the existence of a social production function where the marginal productivity of labour and capital is declining – both individually and together. Consequently, in the long-term natural resource availability will form a limitation to growth. While the position of economists and others holding a more neo-classical viewpoint is that the long-term future is unpredictable, and that supply and demand will be adjusted so as continuously to maximise social welfare in relation to resource availability. Therefore, the marginal productivity of capital and labour will continue to increase.

The Malthusian doctrine lies behind much of the concerns of the conservation movement and its present 'green' successors. As President Taft of the United States said in 1910 'A great many people are in favour of conservation, no matter what it means' (Taft, 1910). The basis of Malthus' thesis, as was that of the conservation movement of the late 19th and early 20th centuries, was the dependence of contemporary industrial society on land and natural resources. The basis of the thesis was the following – never proven – supposition:

> if the natural increase in population, when unchecked by the difficulty of procuring the means of subsistence or other peculiar causes, be such as to continue doubling its numbers in twenty-five years, and if the greatest increase of food which, for a continuance, could possibly take place on a limited territory like our earth in its present state, be at the most only such as would add every twenty-five years an amount equal to its present produce then it is quite clear that a powerful check on the increase of population must be almost constantly in action (Malthus, 1830).

This relationship did not hold at the time Malthus wrote, for agricultural land or for any other resource, and it certainly does not hold for any resource today. It was and remains a hypothesis, one which time has disproved. There are no limits to growth. In all practical terms, the natural resource base is infinite, as demand and supply for any given resource adjust through the market. In fact, the relative scarcity of natural resources has been in continuous decline over the last two centuries. There is, however, the apparent exception of freshwater. For freshwater, the Malthusian argument appears to hold, as no increase in the natural supply is possible. The balance of the hydrological cycle acts to fix the quantity of freshwater available in nature. It follows, therefore, that water, alone among all the natural resources, meets the necessary conditions to support the Malthusian growth model (Barnett, 1960):

1. in the long term freshwater resources will become scarce once population rises beyond a certain critical level;
2. the critical population/resources level will be reached with exponential population growth, whatever the fertility rate, as long as the rate of population growth exceeds the maintenance level;
3. the amount of freshwater is fixed as was Malthus' 'fixed land';
4. the advance of technology and institutional solutions is too slow to offset the decline in marginal returns arising from population growth;
5. total social output will show diminishing marginal returns in respect to the water resource.

For the Malthusian model of growth to apply, however, it is necessary that all five of these conditions exist. If even one of the conditions is not met then the model will not in the long term result in subsistence living and a cessation of economic growth or, as is germane here, in a generalised social crisis caused by water scarcity.

One of the necessary bases for the Malthusian argument to apply is the continuous expansion of the human population. If the supply of water is finite, 'fixed', despite the fact that water is recycled and reused, then as the population increases it is very obvious that the supply per capita must decline. Of course, in reality, we do not use water in this abstract way. We use none of the natural resources that way. To the extent that we inhabit a finite, limited planet then as population increases the per capita supply of all natural resources declines, as long as, in the Malthusian sense, population expands more rapidly than the creation of new resources or in the institutional and technological ability to reduce use per unit of output. One question that must be addressed becomes, therefore, the likely future trends in the growth of human population. Will population continue to increase as rapidly as it has done over the last 10,000 years?

Population Expansion

The human population has increased rapidly and significantly particularly in the last 300 years from around 500 million to an estimated almost 6 billion in the year 2000 (Figure 1.2). Extrapolating these historical rates of increase

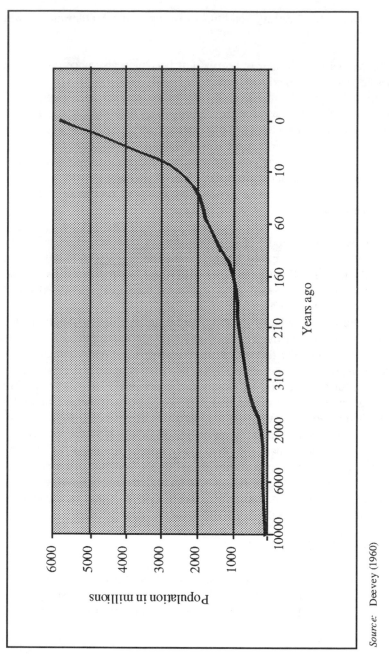

Population in millions

Years ago

Source: Deevey (1960)

Figure 1.2 Expansion of the human population over the last 10,000 years

into the future would suggest that a population-related catastrophe is indeed in the human future, but this may not be the case. The most recent population projections prepared by the United Nations provide an estimate that the world population, based on what is termed a medium- fertility scenario, will stabilise around the end of the 23rd century, at slightly less than 11 billion. This is around twice the level existing in 1995, but a smaller population than estimated in earlier projections, (United Nations, 1998). This almost doubling the present world population means, of course, halving the absolute per capita supply of all resources, including water. On the basis of this projection, however, this would be the end game with a future of human population stability for the world as a whole.

The United Nations, however, does not provide just this one projection, but rather a range of projections based on what it terms different fertility scenarios, the average number of children born to each woman (Table 1.1). The difference between the highest and lowest of these scenarios is only one child per couple, but the resulting populations are markedly different: 3.6 billion to 27 billion persons. It is important and very thought-provoking to note that the low projection would result in a population considerably smaller than the current population. The difficulty with these projections as with all such long-term forecasts, is that we do not, and cannot, know which projection is the most likely to occur. If population were to go on increasing at the fertility rates of the period 1990 to 1995 then by the year 2150, the population would be 296 billion persons! The report suggests, however, that the medium scenario, which assumes that fertility will stabilise at

Table 1.1 World population projections, 1950–2150 (in billions)

Year	High	High/ Medium	Medium	Low/ Medium	Low	Current levels
			Fertility Scenarios			
1950	2.5	2.5	2.5	2.5	2.5	2.5
1995	5.7	5.7	5.7	5.7	5.7	5.7
2050	11.2	10.8	9.4	8.0	7.7	14.9
2100	17.5	14.6	10.4	7.2	5.6	57.2
2150	27.0	18.3	10.8	6.4	3.6	296.3

Source: United Nations, Population Division (1998).

replacement levels of slightly above two children per woman, is the most probable future demographic situation given current trends in fertility.

This assumption, however, is being questioned as not being sufficiently radical in forecasting long-run trends in population. A large number of countries now have fertility rates considerably below the replacement level. The average fertility rate in Europe is 1.5 births per woman, in Hong Kong 1.3 and North America and other Asian countries also currently show fertility rates below the replacement level. Therefore, population decline rather than continued increase may be the situation that many societies will have to confront. This assumption will be incorporated in the new projections currently being prepared by the United Nations (Bongaarts, 1998).

Future growth of population, due to the regional differences in fertility rates, will be accompanied by changes in relative distribution unless there are large migratory movements. It is projected that the population of Europe will decline so that by the year 2150 it will be 18 per cent lower than it was in 1995. The largest increase in population will be in Africa, which will quadruple its population to 2.8 billion. In the Americas and most of Asia population increases will be similar to the world average, a doubling, except in China where the population is only projected to increase by one-third.

So we can conclude that there appears to be significant evidence from the already apparent changes in fertility rates that the human population will not continue to increase as it has done so for so long. This conclusion may be difficult to accept after decades of concern over the expansion of the human population, but accept it we must if we are to make sensible decisions about the future. This element in the Malthusian model may not hold in the future. Changes in demographic growth may make a major contribution to the reconsideration of the impact of population on the future demand for water, as for all other natural resources.

FORECASTING FUTURE WATER DEMANDS

Two prize winning scientists warned on Tuesday that world leaders would have to address highly sensitive political issues in the coming 30 years to avoid bloody wars over scarce water resources *Infobeat News*, (1998)

A recent report published by a group of international organisations, with support from the Government of Sweden, illustrates the importance of the future size of the population in determining the future demand for water. The report provides an evaluation of the likely future impact of population increase on the demands on the world's water resources (WMO, 1997). The report attempts to provide a comprehensive assessment of what is termed future water needs and is the first report of its kind in which an attempt is made to define what water scarcity is. Using estimates of present water withdrawals and water availability, the authors of the report classify the countries of the world by income and by what is termed water stress (Table 1.2). The definition of water stress is the relationship in a country between water availability and water withdrawals for human use. Countries are classified into four groups, low water stress, moderate water stress, medium-high water stress and high water stress. A country suffers from high water stress when more than 40 per cent of the available water is withdrawn for use. The report describes such countries as suffering from serious water scarcity with increasing reliance on desalination and using groundwater at a rate faster than it can be replenished. Water scarcity in such cases, it is claimed, can become a limiting factor for economic growth.

Table 1.2 Distribution of the world's population by income and water withdrawals

Per capita income (US$ 1995)	Withdrawals as per centage of available water				
	<10%	10–20%	20–40%	>40%	Total
< 795	806	1266	958	238	3268
796–2985	542	286	165	138	1132
2896–8956	259	13	137	63	473
> 8957	108	514	181	20	824
	1722	2079	1442	459	5696

Source: World Meteorological Organisation (1997).

In 1995, according to the estimates obtained, a number of countries already could be classified to be suffering from high water stress. These countries were mainly in North Africa and Western Asia, although Belgium and South Korea also fell into this group. The difficulty in interpreting the significance of these results is in the very precarious basis of the estimates used for arriving at the relationship between population and water withdrawals.

Total water withdrawals are not something that is actually measured in any country. Attempts to quantify withdrawals have to depend on very general assumptions about the relationship between say an estimated agricultural acreage under canal and the water withdrawn for irrigation or industrial production in tons and water withdrawn for industry. The use of this type of relationship means that the estimates of withdrawals, in any particular case, are subject to considerable margins of error. In addition, although better measured in some countries, the quantities of available water are not accurately known either. This is true even for streamflows and is much more the case for estimates of the water available in lakes, runoff, soil moisture and groundwater. One can be justified, therefore, in being sceptical about the conclusions of any report which pretends to estimate these quantities and from there to reach conclusions on current and future levels of water scarcity no matter how carefully prepared and cautiously interpreted.

The precarious basis of the estimates for water withdrawals and the dependence of the conclusions on equally statistically precarious population projections are weaknesses admitted by the authors. For the first time, however, in a report published by the United Nations, water management policies are discussed in the context of the realisation that water is a scarce resource and that it should be treated as an economic good. It is difficult, however, to accept that water scarcity will actually act as a brake on future development as it can be expected that the realisation that water is a scarce resource will encourage increasing efficiency in use. The report recognises that this will probably be the case. It goes on to advocate the general introduction of water markets as a means 'to encourage the private sector to play an important role in providing the necessary financial resources and management skills needed for successful development and utilisation of the resources' (WMO, 1997). The argument may be here a little confused. The private sector has not needed water markets to invest in water-related infrastructure, but only clear and stable public policies. Water markets are required to ensure the efficiency of water allocation, with or without private investment.

The danger present in this report, as in much of the debate on water scarcity, lies in the fact that it confounds water requirements and water demand. As was pointed out 40 years ago by Ciriacy-Wantrup when he noted that, although the elasticity of the demand for water is small in respect of price elasticity it tends to be proportional. Proportionality means that in such a case where 'the expected upward movement of water prices starts at low – sometimes zero – levels. Increases will be large proportionally. With such large proportional changes in prices, even small elasticities of water demand – let us say around 0.10 – lead to considerable absolute changes in quantities' (Ciriacy-Wantrup, 1961). Yet requirements or needs are still commonly considered to be the equivalent of demand.

Is the Demand for Water Increasing?

The comprehensive assessment discussed above and other less balanced reports on the future demands to be placed on the water resource depend on two basic assumptions. The first assumption is the continuing expansion of the human population, which has already been discussed, and the second, the increasing demand for water. The assumption that the demand for water will increase with population growth has logic, although it may not hold, but it is also commonly assumed that the demand for water will increase with economic growth. This assumption is not quite so impelling. It has been accepted for some time 'that the proposition that price of services of natural resources must rise relative to the services of reproducible capital over time as a consequence of economic growth is demonstrably false' (Schultz, 1961). It is false because demand for natural resources does not necessarily increase with economic growth. Quite the reverse seems to be the case. It is just one more example of the phenomenon '(B)ad news about population growth, natural resources, and the environment that is based on flimsy evidence or no evidence at all is published widely in the face of contradictory evidence' (Simon, 1980).

The question for the present discussion is then: is there any evidence that would provide any indication of the relationship between the use of water resources and economic growth? Giving an answer to this question is difficult given the discussed precarious nature of the evidence available on the demand for water. Fortunately, however, there is one relatively reliable recent source available. A source, which suggests that for water, as for other resources, the demand does not necessarily increase with economic growth. Consequently, therefore, growth may not result in greater scarcity of the resource, even in the absence of market allocation and real prices.

In 1998, the United States Geological Survey issued a report estimating trends in the use of water in the United States (Solley, Pierce and Perlman,

1998). The report brings together estimates for water use in the United States from 1950 to 1995 (Table 1.3). The results of the new estimates of water use in the United States and of the comparisons made in the report with early years are perhaps a little surprising within the context of the public discussion of a future of increasing water scarcity. In synthesis the report shows that in 1995, the estimated withdrawal of fresh and saline water for offstream uses was 2 per cent less than had been estimated in 1990 and 10 per cent less than the levels estimated for 1980. This decline occurred despite the increase in the population of the United States by 16 per cent between 1980 and 1995 so that per capita withdrawals decreased by an even larger proportion than the decline in total withdrawals. Interestingly, the decline in withdrawals was entirely from groundwater between 1990 and 1995, withdrawals from surface water remaining stable.

There is perhaps one development over the last five years which lends some support to the thesis of a future increasing scarcity of water. Over the period 1990–1995, the consumptive use of water was estimated to have returned to the 1980 level following a 6 per cent decline between 1980 and 1990. Consumptive use, however, only represents a quarter of total withdrawals and the change reflects perhaps, in part, increasing efficiency in the use of water.

These changes in water use in the United States have occurred in a context where, as elsewhere in the world, water is generally allocated outside the market process. In the United States, as elsewhere, there is only rarely a price for water. The user may have to pay the cost of abstraction, but this does not represent a price. It could be hazarded that if the price charged for water in the United States represented the true economic value of water then the decline in per capita withdrawals might be much greater than that shown in the study of the Geological Survey.

The World Bank has been concerned for many years on the result of pricing water below its real value:

> Pricing water well below its economic value is prevalent throughout the world. In many countries, expanding the supply is politically expedient, and therefore pricing and demand management have received much less attention. The preference for expanding supply has led to ... pressure on water dependent ecosystems (World Bank, 1993).

Table 1.3 Estimated water use in the United States, 1950–1995 (thousands of millions of gallons per day)

	1950[1]	1960[2]	1970[3]	1980[4]	1990	1995	Percentage change 1990–95
Population, in millions	150.7	179.3	205.9	229.6	252.3	267.1	+6
Offstream use:							
Total withdrawals	180.0	270.0	370.0	440.0	408.0	402.0	–2
Public supply	14.0	21.0	27.0	34.0	38.5	40.2	+4
Rural domestic and livestock	3.6	3.6	4.5	5.6	7.9	8.9	+13
Irrigation	89.0	110.0	130.0	150.0	137.0	134.0	–2
Thermoelectric power	40.0	100.0	170.0	210.0	195.0	190.0	–3
Other industry	37.0	38.0	47.0	45.0	29.9	29.1	–3
Total consumptive use (freshwater)	nd[5]	61.0	87.0	100.0	94.0	100.0	+6
Instream use:							
Hydroelectric power	1,100.0	2,000.0	2,800.0	3,300.0	3,290.0	3,160.0	–4

Notes:
[1] 48 States and District of Colombia
[2] 50 States and District of Colombia
[3] 50 States and District of Colombia, and Puerto Rico
[4] From 1980, 50 States and District of Colombia, Puerto Rico and Virgin Islands
[5] no data

Source: Solley et al (1998).

17

We can go further than the World bank did in this statement in its water management policy paper. We can claim that the whole idea of a water crisis and of water scarcity acting as an obstacle to continuing human developmentand being the possible cause of future wars arises from the basic failure to correctly price water, in fact, to chronically under-price it. Under-pricing water and the failure to distinguish between requirements and demand when coupled together with the perception of water as different from other resources results in a heady mixture. It leads even serious professionals to draw erroneous conclusions about the possibility of the existence of a water crisis and leads to the mistakenly foreseeing the possibility, if not the probability, that water scarcity will form a serious obstacle to further human development in many parts of the world.

Having said that can we safely conclude that water scarcity is a myth, the figment of the over-active imaginations or the desire for publicity of some commentators? To a large extent, water scarcity is undeniably an end of century myth. However, the possibility of scarcity cannot simply be dismissed. The distribution of water and the distribution of human populations are uneven and do not always coincide. The issue of scarcity must be given due consideration in water management, but should not be exaggerated into crisis and catastrophe. Today, the Jeremiads have now focused on water, the one resource we fail to assign through the market. Water is badly distributed among uses in many parts of the world and this will not be easy or quickly remedied. It is, however, exactly our failure to use the market to price water that is producing these real problems in the distribution of water among uses. The problem must be tackled, but by correctly pricing the resource, not by massive intervention by governments that, in many cases, are incapable of managing the most basic tasks required of them.

REFERENCES

Barnett, Harold J. (1960), 'Malthusianism and conservation: their role as origins of the doctrine of increasing economic scarcity of natural resources', in *Demographic and Economic Change in Developing Countries,* Universities-National Bureau Committee for Economic Research, Special Conference Series No. 11, Princeton, New Jersey: Princeton University Press, pp. 423–456.

Bongaarts, John (1998), 'Demographic consequences of declining fertility', *Science*, **282**, 419–420.

Chonchol, Jacques (1994), *Sistemas agrarios en América Latina*, Mexico City and Santiago, Chile: Fondo de Cultura Económica.

Ciriacy-Wantrup, S.V. (1961), 'Projections of water requirements in the economics of water policy', in H.L. Amoss (ed.), *Water, Measuring and Meeting Future Requirements*, Western Resources Conference, 1960, Boulder, Colorado: University of Colorado Press, pp. 211–226.

Deevey, Edward (1960), 'The human population, *Scientific American*, **203** (9), reprinted in Ian Burton and Robert W. Kates (eds) (1965), *Readings in Resource Management and Conservation*, Chicago: University of Chicago Press, pp. 10–20.

Infobeat News (1998), 28 October.

International Panel on Climate Change (IPCC) (1997), *The Regional Impacts of Climate Change: Summary for Policymakers* (SPM), electronic version

Lean. G. and D. Hinrichsen (1994), *Atlas of the Environment*, New York: Harper Perrenial.

Malthus, Thomas (1830), 'An essay on the principle of population', reprinted in F.W. Nottestein (ed.) (1960), *On Population, Three Essays*, New York: The New American Library, pp. 13–59.

Periera, H.C. (1973), *Land Use and Water Resources*, London: Cambridge University Press.

Schultz, Theodore W. (1961), 'Connections between natural resources and economic growth', in J.J. Spengler (ed.), *Natural Resources and Economic Growth, Conference Papers*, Washington: Resources for the Future, reprinted in Ian Burton and Robert W. Kates (eds), (1965), *Readings in Resource Management and Conservation*, Chicago: University of Chicago Press, pp. 397–403.

Science (1998), 'The next oil crisis looms large – and perhaps close', *Science*, **281**, 1128–1130.

Simon, Julian L. (1980), 'Resources, population, environment: an oversupply of false bad news', *Science*, **208**, 1431–1437.

Solley, Wayne B. Robert R. Pierce and Howard A. Perlman (1998), *Estimated Use of Water in the United States in 1995*, U.S. Geological Survey Circular 1200, Washington: U.S. Department of the Interior.

Taft, William H. (1910), *Outlook*, May 14, p. 57, quoted in Harold J. Barnett (1960), 'Malthusianism and conservation: their role as origins of the doctrine of increasing economic scarcity of natural resources', in *Demographic and Economic Change in Developing Countries,* Universities-National Bureau Committee for Economic Research, Special Conference Series No. 11, Princeton, New Jersey: Princeton University Press, pp. 423–456.

United Nations, Population Division (1998*), World Population Prospects: the 1996 Revision,* New York: United Nations.

White, Gilbert F. (1957), 'A perspective of river basin development', *Law and Contemporary Problems,* **22** (2), 157–184.

White, Gilbert F. (1970), Preface to *Integrated River Basin Development,* 2nd edn, United Nations New York, reprinted in Robert W. Kates and Ian Burton (eds) (1986), *Geography, Resources and Environment,* Chicago: University of Chicago Press, pp. 253–265.

Wittfogel, Karl A. (1957*), Oriental Despotism: a comparative study of total power,* New Haven and London, Yale University Press.

World Bank (1993), *Water Resources Management,* A World Bank Policy Paper, Washington D.C.: The World Bank.

World Meteorological Organisation (1997), *Comprehensive Assessment of the Freshwater Resources of the World,* Geneva: World Meteorological Organisation.

2. Institutional Approaches towards Water Management

Given the ancient and global tradition of water management as a public activity, it would not be surprising to find many alternative ways of undertaking this task. In reality, on closer examination of many water management systems, it is quite clear that this is not the case. Water management, historically, has taken only two general forms, although there are many variations on these. We have already discussed how the strong influences of the autocratic hydraulic societies influenced the whole history of the development of water management and the perception of water as a resource. A centralised top-down approach is then the one major form that water management has taken while the other is the inverse, where management is undertaken from the bottom-up through local users who developed organisations to tackle common problems.

Modern scientific water management practices, which have their beginning in the last half of the 19th century, were for many years identified only with the centralised approach towards water management. This was the form in which the first large modern water projects were conceived and put into execution. However, it was also towards the end of the 19th century that water laws began to be codified and it was also in these years that user-based organisations began to receive formal legal recognition. The latter part of the 19th century was also the period in which concern about the impact of human activity on the natural environment began to become a public issue. These different events converged to create trends and tendencies that have dominated water management in this century.

It is not always convincing to try to establish an order in the historical development of any activity, but there is an observable change in approaches to water management beginning around 150 years ago. From

1850 on there are numerous examples of the beginning of the adoption of more formal methods. To cite a few disparate examples, towards the end of the 19th century the British carried out comprehensive surveys of hydraulic conditions in Egypt and India which led to the construction of large irrigation systems on the Indus and the Nile. Similarly, it was in the years after the Civil War that the conservation movement emerged in the United States, in Prussia (Germany) the foundations of the *Genosschaften* were laid and in Chile, the Civil Code of 1851 established the basis for private sector participation in water management, particularly in irrigation.

All these developments, wherever they occurred, were no doubt, in part, a reflection of the path of the general economic and social developments over the last 150 years. Scientific approaches to water management reflect the 20th century concept of the application of technology to the solution of economic and social problems. The survival and normalisation of the role of user organisations in water management reflect the strong traditions of local government existing in many countries. The influence of the conservation movement within water management is a reflection of the overall growth in concern for the quality of the environment.

Over most of 20th century, however, it was the centralised top-down system of water management, which became identified with scientific management. It became the preferred form of water management and dominated approaches towards water management in most countries. The bottom-up user managed organisations did not disappear, but for most of the 20th century they have been very much marginal to the main trends in water management. Gradually the state occupied most water management space until its abrupt retreat in the middle of the 1980s (Figure 2.1). Under the influence of socialist ideas, public sector investment and the role of government in the economy as a whole grew continually until very recently in nearly all countries, both developed and developing. As part of the general expansion of the public sector and of the role of the government in the economy, public sector investments in the control and regulation of river flows expanded. The form of state intervention changed with local single-purpose institutions for water supply, irrigation or power being replaced by national ones. The development of national institutions was accompanied by the increasing creation of multi-purpose water management agencies, normally on a regional, basis but, exceptionally, as in the case of the *Secretaría de Recursos Hídraulicos* in Mexico, national.

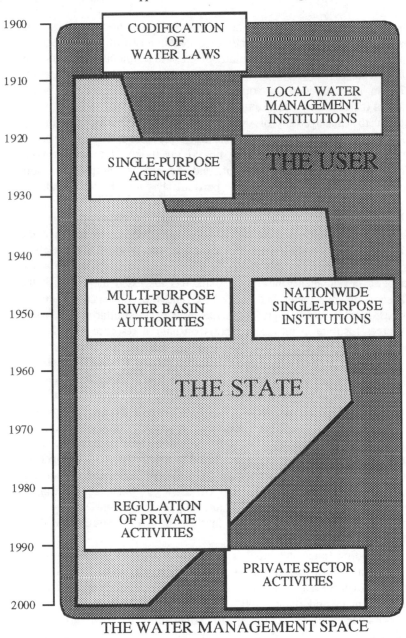

Figure 2.1 Approaches and actions in water management in the 20th Century

It became increasingly obvious, however, that much of this investment was not well directed and projects were commonly failing to meet the objectives set and the returns on many of these investments were very low if not negative. Beyond the confines of water management, since the 1970s in the economic development literature, the role of the public sector in the economy has been reconsidered. In particular the role of the state in the management of directly productive activities and in the provision of services has been increasingly questioned. The emphasis in the development literature is now placed on the need to increase the effectiveness of the management of, and the rate of return from, both existing and future investments, particularly, investments made in projects in the public sector.

In the context of this discussion, not only governments, but also international institutions have shown increasing concern to improve the efficiency of water management. For example, the World Bank has given considerable attention to this issue. In its recent annual reports on world economic development, the World Bank has discussed the issue of the efficiency of investment and the effectiveness of management in the public sector at some length (Frederiksen, 1992). Similarly, the World Resources Institute in its recent reports on natural resource management has emphasised the importance of improving management capabilities to make existing investments more profitable (World Resources Institute, 1993). The result of this reconsideration of the role of the government and the public sector in the economy has been a widespread revival in the faith of markets and the private sector. This turn towards the market has been felt within water management and water resource administration, as in other areas of the economy. The most common signs of this change are the sale or concession of many water-based productive activities to private companies. An even more important institutional change, however, has been the strengthening of the role of users in water administration, particularly in regions of irrigation farming. A few countries have even made water rights into real property that can be freely transferred without reference to the bureaucracy.

For the first time, only now is it widely recognised that users should be given a serious role in water management and local organisations should be considered as capable of applying scientific approaches to management (World Bank, 1993). At last, the bottom-up approach to water management is beginning to come into its own as private participation increases and water is being seen to be an economic commodity. Increasing user responsibilities in water management coincide with the general adjustment in the role central governments play both in the economy and in the administration of public activities.

SCIENTIFIC WATER MANAGEMENT FOR ECONOMIC DEVELOPMENT

One of the major arguments used to justify the expansion of the state role in water management during this century was the desire to make deliberate use of water-related projects to achieve the wider objectives of economic and social development. The basis of this approach to water management has been succinctly defined by Gilbert White in his classic paper of geographic contributions to river basin development, '(B)y river development I mean the control and use of water for multiple human purposes in river basins...' (White, 1963). Gilbert White, along with many contemporaries, saw the development of water management in the 20th century culminating in the application of research as a management tool and the adoption of what he termed a water management strategy of 'multiple purposes and multiple means' (White, 1969).

The control and use of water for the wider aims of economic and social development implied the planned investigation of the resources in a river basin, the identification of the possible alternative uses for these resources, the selection of the best alternatives given available technology, all within the economic and social objectives set for future development. If this approach smacks of socialist-type central planning, it was indeed an approach widely adopted in the Soviet Union and other socialist countries. It was an approach widely used also in India and many countries of Latin America influenced by the Soviet example in the 1950s and 1960s. The origins of this model of water management clearly go back to the great hydraulic civilisations. Its modern revival and application, however, owes much to the operations of the British Colonial Service and of the Bureau of Reclamation and other federal government agencies in the United States and their combined influence within the multilateral lending agencies.

There are many examples of attempts to develop multi-purpose river basin development agencies and projects imitating the famous example set by the Tennessee Valley Authority in the United States. It is not possible to describe them all even briefly here. One interesting example to illustrate the rationale and objectives of this approach is provided by the developments that were proposed, but not carried out, on the Lower Mekong River.

The proposed development of the lower Mekong River Basin

The project to develop the drainage basin of the Mekong below the Lao–Burma–Chinese border where its gorge cuts deep through the rugged Yunnan Plateau is an excellent demonstration of an attempt at integrated river basin development. The Mekong Committee was founded in 1957 and became the focus of a major international effort to use water management to achieve economic and social development in the early 1960s. Due to the prolonged wars in Vietnam and Cambodia, however, the efforts were seriously interrupted and then abandoned in their original form and scope. The Mekong Committee still exists, but Cambodia withdrew in 1975, and the committee now largely restricts its activities to the collection of hydrologic data (IHGE, 1995).

The Lower Mekong basin covers an area of about 600,000 square kilometres in Vietnam, Laos and Cambodia with a population of more than 22 million with relatively low incomes, mainly generated from agriculture and forestry. In the early 1960s, wet-season rice was the principal product. Upland rice was grown in the burned clearings of shifting agriculture and paddy in the valleys of the plateaux and in the broader alluvial plains that stretch from below the Sambor Falls and around the Great Lake (Tonlé Sap) to the lower margins of the Delta. Before the project, despite the importance of the river to the local societies, water management outside the individual paddy fields was of minor significance.

The undertaking of this massive international effort was justified on the basis that the development of water-related projects was one of the most promising means of supporting economic growth in the basin countries. It was estimated that if, over the 25 years after 1960, agricultural production was to meet demands of a population growing at 3 per cent per year, and at the same time lead to exports and support a modest increase in incomes, a strong expansion in agriculture and in energy production linked with industrialisation would be necessary. The river has sufficient flow and gradient to generate significant electrical energy and to supply irrigation for dry season cultivation and for opening up new lands. Flood control and low-water control were considered essential in many sections for the improvement of crop production by irrigation and drainage. Channel improvements would promote cheap water transport along the Mekong.

The volumes of water, the hydraulic heads, the availability of lands, and the readiness of governments to join in co-operative studies combined, it was claimed, to favour a comprehensive approach to water management.

Table 2.1 The Mekong project: summary of the investigations

Activity	Country or Agency	Estimated Cost (US$ 1960)
Mapping	Canada	1,365,000
Topography	Philippines	235,000
Levelling	Canada and USA.	a
Hydrographic surveys and	New Zealand and U K	183,000
navigation studies	United Nations	364,000
Hydrologic measurements	USA	2,200,000
	WMO	45,000
Dam site explorations	Australia	409,000
Soil surveys	France	a
Minerals prospecting	France	750,000
	United Nations	a
Fisheries studies	France	
Major engineering projects:		
Pa Mong	USA	2,500,000
Sambor	Japan	a
Tonlé Sap	India	282,000
7 Engineering		
reconnaissance projects	Japan	652,000
	United Nations	2,719,000
	Israel	75,000
	Pakistan	b
Agricultural pilot projects	United Nations and FAO	a
Public health study	WMO	5,000
Delta model	UNESCO	17,000
Shifting agriculture/forestry	FAO	186,000
Manpower study	ILO	12,000
Economic and social studies	Ford Foundation	b
Flood warning system	France	a
Power market surveys	France	a
	United Nations	a
	Resources for the Future	b

Notes:
    a. Cost included under other item for same agency.
    b. No cost estimates available.
Source: White (1963).

It was admitted, however, that there was a critical lack of data, a shortage of trained personnel, and little capital. For example, in 1957 there was not a single long-term accurate measurement of discharge of the river, and population estimates for some areas of the basin varied by as much as 40 per cent. In order to undertake the project a co-operative effort was organised by a consortium of 14 countries, 11 international agencies, and several private foundations (Table 2.1). At the time, it was the largest project of it type ever undertaken.

A large international investigation of the physical characteristics of the basin was the first major task. This included the production of large-scale maps of promising dam sites and reservoir areas, geological investigation of dam sites and of possible mineral bodies suitable for commercial exploitation. Engineering studies were undertaken for three projects on the Mekong and seven on tributaries, including hydrology and soil studies as well as design work. A skeleton network of hydrologic observations was established among other hydrologic studies.

Concern was expressed by those involved in the project that final design and construction of any river project on the Lower Mekong could incur heavy costs in the form of projects that failed to achieve their anticipated social gains. It was also pointed out that designing effective schemes for agricultural settlement was very much more difficult than designing and building a dam. The recommendations to overcome the possible failure of the project to achieve its objectives were, however, to make even more expensive studies (Table 2.2).

White, in commenting on these additional studies to be made by social scientists – economists, anthropologists, administrators, and geographers – said:

> Implicit in these recommendations is the belief that there is a strategy or alternative set of strategies for water management that will better serve the welfare aims of the four Lower Mekong countries than others. The kinds of strategies which they might adopt and the ways in which they might make their selection are highly diverse, as the record of efforts at integrated river development around the world reveals most colorfully (White, 1963).

An optimum strategy perhaps existed. However, in their determination to improve the life of the inhabitants of the Lower Mekong Valley, none of the authors of the project seems to have considered that the first step should not have been to arrive from the moon, as it were, and study the situation and then decide to build this or that project. The project managers seem to have taken no or little account of the previous absence of any water management institutions, which could form a basis on which to construct the gigantic developments proposed. It was forgotten, or rather not considered, that what had to be done was the far more difficult task: the task of creating the conditions so the population of the four countries, themselves, could decide on their own management of the water. It is highly likely that the reply would have been that the governments wanted the projects and had requested the studies and the support of the international community. There is, however, a need, in responding to such a request, to consider the real capabilities existing in the society in which it is to be mounted for managing the project.

No one at the time, in commenting on the failure of these ambitious efforts, seems to have realised that what was wrong was the scale. It was the attempt to build – in the Mekong valley, in the Tennessee Valley, in the São Francisco valley in Brazil or in the Damodar valley in India – systems of water management independently of existing administrative institutional structures. These systems were far more elaborate and complex than could be managed locally and, in the specific case of the Lower Mekong, even nationally. The consequence in all cases, even where successful water control structures were built, was failure and waste, in terms of the original objectives set for economic and social development. It is abundantly clear that:

> the record of multipurpose river development in many parts of the world is unhappily ornamented by instances where less than the full range of possible uses of water management was considered, often with serious social consequences. …ineffective efforts at water development have come from people who set out to implant the methods of another culture in inhospitable social soil (White, 1963).

There is, in addition to the problem of scale, the fact that multipurpose projects, despite their popularity among the development and water management bureaucracies, were bound to fail due to their complexity. The sheer size and complexity of these efforts ensured that they were beyond the possibilities of any institutional organisation to manage. The fact that any water body can have multiple uses should not have determined that all developments should be multiple in the face of the limitations of administrative and institutional reality.

In many countries, there has been a tradition of heavy dependence on centralised command and control administrations for developing and managing water resources and excessive reliance on government agencies to develop, operate and maintain water systems. In many instances, this has stretched the already limited implementation capacity too thinly.

Table 2.2 The Mekong project: recommended additional studies

Problem	Method of work	Approximate Cost (US$, 1960)
Synthesis of available data on resources, resource use, and social characteristics	Basin atlas	180,000
Inventory of land use and land-use capability	Aerial photography and mapping	2,500,000– 6,500,000
Assessment of agricultural improvement measures	Study and seminar	135,000
Comprehensive analysis of power market potential	Country surveys Regional market study World market appraisal	120,000– 600,000 150,000– 450,000 80,000
Flood forecasting and damage reduction	Analysis of present and possible adjustments to flood hazard	45,000
Examination of methods of estimating economic feasibility	Joint review in field and in office by study groups	100,000
Scale and scope of ultimate Mekong system	Simulated systems analysis	225,000
Study of critical problems of economic and social analysis	Research and short-term training	500,000
Study of regional problems and prospects	Co-operative study under international agencies	–
Exploration of practical administrative arrangements for international construction and operation	Consultant and subcommittee	100,000
Comprehensive rural demonstration project of 3,000–5,000 hectares under water management	Unified regional administration	2,500,000

Note: The mission also recommended measures to strengthen committee staff and to experiment with forest plantings.
Source: White (1963)

Moreover, in most cases, users have not been consulted or otherwise involved in the planning and management of the water resources. The result has been a vicious cycle of unreliable projects that produce services that do not meet consumers' needs and for which they are unwilling to pay (World Bank, 1993).

The multi-purpose approach towards water management failed because what was usually being proposed was beyond the possible capacity of the public administration to manage and out of scale with the real needs of the potential users. At last, it is becoming recognised and understood that the centralised and autocratic views of water management embodied in this approach are not the optimum means of managing the water resource.

USER-MANAGEMENT APPROACHES TO WATER MANAGEMENT

Many of the examples of successful user-managed institutions are found in societies where irrigation is dominant as in one of the two examples presented here, from the Elqui Valley in Chile. User-managed institutions are not, however, limited to irrigation. In Chile, the user institutions, although dominated by irrigators, include all uses of water. One of the oldest and most successful examples of user-managed institutions is not at all related to irrigation. They are the water management institutions of the Ruhr Valley, the old industrial heart of Germany.

The *Genossenschaften, Ruhrverbund* and *Ruhrtalsperren-verein* in Germany

In the Federal Republic of Germany, water quality management is a shared responsibility between the states and the federal government. The federal government has taken increasing responsibility since the late 1970s for the so-called framework laws which establish minimum common standards, but procedural and organisational regulations remain the responsibility of the states (Germany, 1998). In most states, the responsibility for implementation has been traditionally delegated to lower-level administrative authorities. The development of water management institutions in the Ruhr Valley in Nordrhein-Westfalen provides a good illustration of how user-based organisations can effectively tackle water management issues.

The *Genossenschaften* were originally established before the First World War when the Ruhr was part of the Kingdom of Prussia. In total, nine water resources authorities were established under specific laws between 1904 and 1958 in Germany. All of them are located in what was the Kingdom of Prussia. Since 1937, however, a Federal law exists to regulate the establishment of water agencies across the country (*Wasser-verbandverordnung*) and all the other authorities have been created under this law (Kühner and Bower, 1981). Each *Genossenschaft* is independent and takes the form of an association of private and public entities. These associations accomplish their tasks, the control of pollution and general water management, in co-operation and co-determination with all public and private corporations and other persons who contribute to pollution and who would benefit from any improvements (Fair, 1961).

The *Ruhrverband* was created as a public corporation by the Prussian legislature on 5 June 1913, under the Ruhr River Pollution Control Law. On the same day, the Ruhr Reservoir Association, (*Ruhrtalsperrenverein*), which had existed since 1899, was also made a public corporation by the Prussian legislature by passing the Ruhr Reservoir Law. Both agencies were awarded the right to request mandatory contributions from their members, and their jurisdictions were made equal to the watershed of the Ruhr River.

The particular purpose of the *Ruhrverband* is to maintain good water quality in the Ruhr River and its tributaries. Thus, the *Ruhrverband* has 'to build, maintain, and operate' the facilities that are required to prevent pollution of the Ruhr and its tributaries by the individual members of the association. The association is only obligated to undertake water quality measures beyond those stipulated by the state water laws when severe circumstances cannot be remedied in any other way. The kind and number of required facilities, as well as any alterations and extensions, are subject to the approval of the Nordrhein-Westfalen Minister of Nutrition, Agriculture, and Forestry. The minister can and does delegate this authority to the local administrative district.

The *Ruhrverband* is also authorised to build, maintain and operate, on behalf of its members, facilities which are not required for achieving the association's purpose but which are related to it. This applies in particular to treatment plants serving the special purposes of individual members. The individual member served must pay the costs of these plants.

The members of the *Ruhrverband* are:

1. the owners of mines and of other commercial or industrial enterprises, of railroads, and of other installations that cause pollution of the Ruhr River system or that benefit from *Ruhrverband* facilities;
2. the communities that lie partly or entirely within the drainage basin;
3. the *Ruhrtalsperrenverein*; and, a more recent addition,
4. military camps.

Thus, besides the *Ruhrtalsperrenverein*, both direct and indirect dischargers, industries, commercial establishments and institutions discharging into municipal sewer systems, make up the membership of the *Ruhrverband*.

The job of the *Ruhrtalsperrenverein* is to replace the water that has been withdrawn from the Ruhr River system and not discharged back and to improve the utilisation of the hydropower potential of the river system. To this end the *Ruhrtalsperrenverein*:

1. builds and operates reservoirs;
2. supports the construction and operation of reservoirs by others;
3. builds and operates works to transfer water from the Rhine River and any other necessary facilities.

The members are the water supply companies, and other companies withdrawing groundwater and surface water from the Ruhr River and its tributaries, who withdraw more than 30,000 cubic metres per year, and the users of the Ruhr River system's hydropower. Companies and municipalities may be members of both the *Ruhrverband* and the *Ruhrtalsperrenverein*, if they both discharge effluent and are self-suppliers of drinking water or water for industries. In 1938, the staffs of the two organisations were merged and a single directorate was created. However, they are still two separate legal entities, with separate boards of directors.

Industrial and commercial establishments only become members if their financial contributions would exceed 1/100,000 of the year's total required contribution. The number of members has been declining as industrial restructuring has reduced the number of smaller industrial companies. For example, between 1970 and 1977 membership fell from 1,229 to 991 (63 communities and 928 industrial, commercial, and other activities). Suggestions to raise the minimum contribution for membership have been rejected on the grounds that such a proposal would exclude the very group, many of them basement electroplating operations or their equivalent, which can cause severe damage to the basin's smaller streams if excluded. Further, with a low minimum figure there is less chance for business firms to feel unjustly treated because a slightly smaller competitor does not have to make payments to the *Ruhrverband*.

In general, municipal treatment plants are *Ruhrverband* facilities. This implies a sharing of construction and operation costs of the sewers and of a contribution to the *Ruhrverband* for wastewater treatment. Therefore, the *Ruhrverband* has many options from which to choose the regional scope and type of treatment system. If a municipality cannot economically connect areas under its jurisdiction to the regional sewer system, and if the *Ruhrverband* does not agree to build a separate treatment plant, the municipality is nevertheless obliged to build its own, in order to meet the effluent requirements established by the state government.

The *Ruhrverband* has three administrative organs, as do all the associations, an assembly, a board of directors and a board of appeals (Fair, 1961). The assembly is the governing body and representation is proportional to the annual contribution, one member for each 1 percent of the annual budget. Smaller contributors pool their contributions to elect members. The Assembly elects the Board of Directors, approves the budget and chooses the members of the Board of Appeals, although the state government nominates some members. The Assembly normally meets biennially. In between meetings the Board of Directors runs the association. The assessments of the association have the same legal status as tax contributions and objections to the assessment are heard first by the directors then referred to the Board of Appeal. Originally, its decision was final, but increasingly appeal is now made to state courts.

The river basin associations in the Ruhr have existed successfully for close to a century despite the changing demands placed upon them and the tremendous change in state and federal laws that affect their operations. 'They continue to flourish as autonomous, pluralistic organisations with representation of both public and private interests', said Gordon Fair in 1961, and that remains true almost 40 years later.

Asociaciones de Canalistas and *Juntas de Vigilancia* in Chile

In Chile, it is the holders of water rights through their organisations, the *Asociaciones de Canalistas* (user associations) and the *Juntas de Vigilancia* (regulating committees), who manage the allocation of water and are responsible for the construction, operation and maintenance of the water control and distribution systems. These private institutions are completely autonomous and self-financing. The origin of the present system is very old and these institutions have played the leading role in water management and in the development of the irrigation and draining infrastructure for the last 150 years. Currently, these organisations are the owners of the greater part of the water infrastructure, including reservoirs and dams. Projects built with public financing are transferred to them for operation and maintenance and no structures are built with public funds unless the users are willing to finance the larger share of the costs. The public sector, in general, has a subsidiary role in support of these user organisations through financial assistance to private investments and programmes to strengthen their actions.

Under the Water Law of 1981, three types of water user organisations are defined:

1. *Comunidades de Aguas* (water communities) which exist *de facto* whenever there is more than one user of any water body. These communities can be formalised;
2. *Asociaciones de Canalistas* (user associations) which manage irrigation canals, drainage works or water, in general. These associations must be formally constituted and are governed by elected boards of directors. The formation of an association must be approved by the Ministry of Public Works;
3. *Juntas de Vigilancia* (regulating committees) are the highest level of user organisation. They are made up from the associations of users of the canals under their jurisdiction and of other direct users of the water and are responsible for the administration of all water uses in major water bodies or large reservoirs.

All these organisations are administered by elected directors. Voting rights are assigned on the basis of the number of water rights held . The law assigns to these user organisations the responsibility to regulate and administer the water resources and related infrastructure under their respective jurisdictions.

Water Management in the Elqui Valley

The Elqui Valley in northern Chile is one of the sunniest places on earth. It is the site of three of the world's major astronomical observatories due to the clearness of the skies. A corollary is the advantages of the valley for agriculture, particularly the production of grapes and other fruit under irrigation. Agriculture uses more than 80 percent of the water of the valley. The hydrological system of the Elqui River and its tributaries is managed by two regulating committees, the *Junta de Vigilancia del Estero Derecho* for part of the Claro River, a major tributary, and the other, the *Junta de Vigilancia del Río Elqui y sus Afluentes*, for the remainder of the main valley. This is the only example in Chile of a regulating committee corresponding to almost an entire river basin (this description is based on Werner, 1997).

This valley is divided for management purposes into three sections

(Figure 2.2). The reason for this division is that it is the easiest way to manage the problem of return flows by restricting the transfer of water rights to within each section. The result of this regulation is that water users in the coastal area cannot buy water rights on the Turbio River.

A specific share of the water flow is guaranteed to each section, which is expressed in the number of water rights. For the first and third sections, these are about 12,000 rights and for the second section only 1,000. The consequences of the prohibition against exchanging rights between sections is that in the lower reaches there is more competition for water rights than up river. This is because the water demands are greater due, in part, to competition from the expanding recreational centres of La Serena and Coquimbo.

The regulating committee for the Elqui River is managed by a board of seven directors, who take decisions by majority vote. The directors are elected according to the number of water rights held by each user association. Because of the large number of rights in the first and third sections, they each have three directors, while the second section has only one.

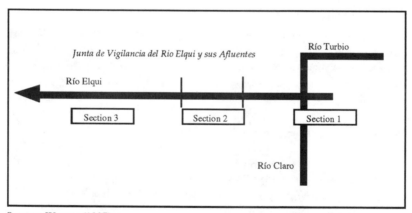

Source: Werner (1997).

Figure 2.2   The user-management system on the Elqui River

The directors make decisions about the acceptance of water transfers, the building of dams, and the annual fixing of the quota of flow per right. The share of each right is fixed on the 30th of April of each year. If this turns out to be too big, then it can happen that down river there is not enough water for all the rights. The exact measurement of the water is very important, because the whole of the flow of the Elqui River is distributed to the users.

In total, there are 115 canals, 85 in the first sector, 9 in the second and 21 in the third. Most of these canals were built in the second half of the 19th century. There are another 25 canals administered by the *Junta de Vigilancia del Estero Derecho*. Each canal has its own user association.

Once the regulating committees have decided on the effective water volume per water right for the coming year, this is fixed for each canal at the intake. A measuring gauge is installed, on which the allocation for the current year is noted.

The holders of water rights pay for the operating costs of the regulating committees and of their canal associations. The contribution of each association to the regulating committee and of each water user to the association is determined according to the number of water rights. On the Elqui River and its tributaries, this administration charge for each full water right (1 cubic litre a second) is at present on an average about US$ 60 per month.

The major problem facing the users is the provision of an assured supply, which arises not from the average amount of water available, but from the unevenness in the flow. Therefore, the building of a big dam has been under discussion for a long time. Until now, the only reservoir in the Elqui Valley, which can be regulated, is the 40-million m^3 La Laguna Reservoir. It does not provide sufficient storage even to meet a drought period of only one year in length. It was decided in March 1996 to build a reservoir of over 200-million m^3 capacity, which will be completed at the end of 1999. It is hoped to guarantee a permanent water delivery, ever more important with the increasing investment in grapes and other permanent crops. The decision to build the dam was taken, not to increase the area under irrigation, but to increase the guarantee of flow. The users are obliged to reimburse the government 60 per cent of the costs of building the dam over a 25 year period.

In both the Elqui Valley in Chile and the Ruhr Valley in Germany, there are complex and sophisticated physical systems for water control and management. In both cases, the water users have largely created these systems. These are multi-purpose and integrated systems, but they were not conceived as whole units and were not imposed on the local population from without by some higher authority. The systems have developed by marginal increments to meet changing demands placed on them by their managers and owners, the local users. Even the brief descriptions given here show clearly that, in both cases, the institutions for water management are sustainable, viable, effective and efficient. They are splendid examples of what can be achieved by local, user-dominated water management institutions.

THE RIVER BASIN AND WATER MANAGEMENT

Whether the organisations managing water are large centralised bureaucracies or user-dominated they operate in a world where water flows downhill, generally to the sea, through a hierarchy of watersheds. This indisputable fact has led many experts on water management to advocate the use of institutions based on the river basin as the logical manner in which to manage water. This opinion, however, has not found many lasting echoes in governments or in society as a whole. In general, and there are very few exceptions, water management institutions, whether national, regional or local, tend to respect administrative boundaries rather than watersheds. In discussing institutional approaches to water management, it would seem appropriate to explore why this should continue to be the case and to ask if there are any essential reasons for a watershed or river basin approach to water allocation.

In his classic work on the legal aspects of river basin management, Teclaff comments that 'under the influence of the areal and functional characteristics of drainage the basin has emerged or tended to emerge at different periods of history as a legal entity' (Teclaff, 1967). There has not been, however, any period when the institutions defined in terms of river basin boundaries have dominated water management. River basin agencies, when established, have tended to be ephemeral or to have their functions restricted to very particular areas of water management, commonly the determination of the release schedules in basins with numerous reservoirs or the conservation of small watersheds.

White, in discussing the perspective for integrated river basin development in the 1950s, commented that it seems to consist of three associated ideas, multiple-purpose storage, basin-wide programmes and comprehensive regional development. In discussing these ideas he further comments that comprehensive regional development has not been realised anywhere. Moreover to take such a broad view of water management is to enter an area of very difficult definition:

> to take this broader view of river basin development is to expand water
> resources planning to encompass all aspects of natural resources as related
> to economic growth, including the cultural conditions of the society, and it
> will be necessary to ask whether or not there is a viable line which may be

drawn between one and the other. For if there is no viable line, the attempts to carry water resources analysis beyond the traditional concepts of multiple-purpose, basin-wide development must inevitably lead to comprehensive schemes in which water, in many instances, would play a secondary role (White, 1957).

This is, of course, the dilemma, for no one has been able to draw a viable line and the bulk of the literature discussing the concept of river basin management tends to be nebulous and contradictory when defining its purpose and goals. Not surprisingly, therefore, attempts to apply river basin management in practice have tended to fail. It is not surprising, therefore, that many governments have abandoned policies adopted to apply the river basin concept to water management, as did the federal government of the United States in 1982 and the government of the United Kingdom in 1989. The abandonment of the river basin institution in the United Kingdom came after a quarter of a century of experience of attempting to apply the concept nationally. Twenty-nine river authorities were established as the basic unit for water management under the 1963 Water Resources Act hailed as 'a giant step forward in England's response to contemporary water problems' (Crane, 1969). The establishment of the authorities did not, however, resolve the major issues in water management in England and they were replaced by 10 regional water authorities, again widely recognised as 'a bold and imaginative attack on existing problems' (Day *et al.*, 1986). They turned out, however, to be equally ineffective in resolving the major water management issues in the country, the adequate financing of water supply and sewerage and control of water quality. Under the Water Act of 1989, the regional authorities were shorn of wider water management responsibilities and replaced by water service companies, which were then privatised (Lee and Jouravlev, 1997).

The more recent discussion of water management institutions places the emphasis firmly on the necessity to improve the efficiency of water allocation and water use. In this context there seems to be little that the concept of water management institutions based on the river basin can offer. The failures of initiatives such as the project on the Lower Mekong, and the obvious limited successes of the more ambitious regional-development inspired institutions such as the Tennessee Valley Authority and its imitators, have resulted in a serious questioning of the idea that basin-wide organisations are required for better water management. It is now widely recognised that effective water management does not need basin institutions, which have shown themselves to be inflexible, inefficient and inequitable, but needs improvements in the basis on which management decisions are made (World Bank, 1993). As Frederick concludes in discussing international water conflicts:

> (M)ore flexible allocation mechanisms and efficient water management practices are critical for avoiding future conflicts over international supplies and curbing the rise in water costs. Introducing markets and market-based prices, ...might help promote a more efficient and flexible allocation of water resources located in international basins (Frederick, 1996).

Advocating the introduction of water markets and prices is not to suggest that the natural characteristics of the river basin should be ignored in water management. It is rather to recognise that the simple fact that water flows downhill does not, of itself, justify the establishment of river basin agencies as major institutional players. Even advocates of integrated river basin planning have become more cautious as to where such institutions may be justified, 'Where they (watersheds) coincide with economic, political and institutional boundaries, they may produce a useful delineation for planning purposes; where they do not other institutional arrangements might be more efficient' (Schramm, 1980). There is no longer a consensus for Teclaff's conclusion that 'the basin will remain a legal entity in the foreseeable future' (Teclaff, 1967).

CONSIDERATIONS IN INSTITUTIONAL DESIGN

The justification for government intervention in any economic activity is to correct market failure and to restore or to substitute the requisite conditions for economic efficiency. Unfortunately, government interventions are often non-optimal, in the sense that they fail to either correct market failure, and thereby restore efficiency, or actually introduce additional distortions. Moreover, markets often find ways to mitigate their own failures.

The particular justification for government involvement in water management is based on the several characteristics of the water sector that can result in market failure. These have been summarised in the World Bank policy paper on water management (World Bank, 1993), as follows:

1. The large and lumpy nature of capital investments and the economies of scale tend to create natural monopolies warranting regulation.

2. The relatively large size of water investments and the potential for political interference reduce the incentives for private investment.
3. Uses of water are interdependent and government regulation is required to ensure that users abide by the rules of the game for water allocation and the maintenance of water quality.
4. Some water-related products, such as flood control and control of waterborne diseases, are, at least locally, public goods, which it is difficult to charge for on an individual basis. Therefore, public intervention may be necessary to ensure appropriate levels of investment.
5. Water resources are often developed for use because of their assumed importance for overall economic development.

Public interest theory holds that government intervention protects the public from the abuses of market imperfections listed above (Phillips, 1993). It views the government and the public sector as omniscient and benevolent maximisers of social welfare, that is they attempt to maximise some kind of social welfare function. The theory emphasises government's role in correcting for market failures (Laffont and Tirole, 1991). The large-scale interventions by governments in water management through multi-purpose projects had just this intention. The market was failing through its inability to take advantage of the synergics multiple-purpose development was supposed to provide. Experience has shown, however, that only rarely have these objectives of government intervention been achieved.

Given the different forms of government intervention that have occurred in recent decades, in general, and the state of water-related services under public sector provision, in particular, it does not seem appropriate to assume that governments are sufficiently efficient, fair and wise to be capable of adopting the optimal intervention prescribed by public-interest theory. Moreover, it can be easily observed that the results 'of economic regulation often differ considerably from the predictions of "public interest models" (Joskow and Rose, 1989). Obviously any government intervention will have political as well as economic dimensions. It is now recognised that 'regulation and regulatory processes respond to complex interactions among interests groups that stand to benefit or lose from various types of government intervention' (Joskow and Rose, 1989).

The recognition of government or regulatory failure challenges the assumption that government is a disinterested champion of the public interest. It pretends to explain the pattern of government intervention in the economy in general and water-related activities in particular, emphasising the role that rent-seeking behaviour, interest groups and capture play in the formulation and conduct of government policies with respect to economic matters. Under these arguments the public sector is seen as either unwilling or unable to serve the public interest: unwilling, because it might seek objectives of its own, separate from the priorities of citizens, and unable, because it operates in an environment full of information asymmetries and under budgetary constraint.

The public choice theory views bureaucrats as seeking power and resources by increasing the scope and scale of government intervention beyond the extent justified by the interests of welfare maximisation (Holtram and Kay, 1994). Political business-cycle theory, on the other hand, assumes that political parties are economic institutions which seek the support of electorates badly informed about the motives of politicians and about the conditions in the economy at large. The capture or interest-group theory draws attention to the role of interest groups in the formulation of public policy (Laffont and Tirole, 1991).

The existence of government or regulatory failure means that the identification of market failure is only a necessary, but not a sufficient condition, for government intervention in water resource management or any other area. To be sufficient, the potential costs of government failure associated with intervention must be less than the expected costs of market failure, that is governments must be able to do better than markets. If this second condition is not met, government intervention will impair rather than restore economic efficiency.

Rent-Seeking from Public Intervention and Regulatory Capture

Two broad forms of government failure can be usefully distinguished in discussing optimising the form of government intervention in the provision of public services. One cause is rent-seeking or the pursuit of self-interest by politicians, public sector employees, and other interest groups. The other is regulatory failure or capture which occurs when a public authority falls under the undue influence of some special interest group, whether public or private.

There are two diametrically opposite views on how to explain the apparent deficiencies in public sector management of public services. The traditional view identifies the public sector with the pursuit of the public good and attributes the generally poor financial and production record of most state-owned enterprises and their failure to provide decent quality service to a variety of financial, administrative and managerial problems.

Many early critics of the obvious shortcomings shown by multi-purpose water management took this approach to its improvement. The work of the Harvard Water Group is perhaps the classic example of this view applied to water management (Maass *et al.*, 1961).

An alternative view, in line with the public choice theory, attributes deficiencies in the public sector to the existence of the private objectives of politicians and bureaucrats that divert public sector companies from their stated objectives (Wirl, 1992). The traditional approach does not or only rarely acknowledges that politicians and bureaucrats may be pursuing other objectives than the maximisation of social welfare through the delivery of water-related services. These other informal objectives, or rent-seeking, can have, however, important and pervasive implications for how water systems are actually designed, built, operated and maintained (Lovei and Whittington, 1991).

In the water sector, the potential rent-seekers are numerous. The first, and main, group consists of politicians at all levels of government. Political parties have high costs of operation, high costs of maintaining their organisation and of competing in elections. They maintain this organisation and their electoral appeal by the performance of services to potential voters at all times, not just before elections (Stigler, 1971). One part of the costs of services and organisation are borne by putting a part of the party's sympathisers on the public payroll. By doing so, politicians impose a small marginal cost on each individual taxpayer to repay the elected official's supporters (Lambert, Dichev and Raffiee, 1993). This can result in significantly inefficient use of labour in publicly controlled water projects and in publicly owned water-related utilities.

Another part of the costs of political parties' organisation and operation can be recovered through the sale of regulation (Stigler, 1971). The objective of any regulated industry is to maximise profits. An industry always wishes, therefore, to be confronted with the weakest regulatory constraints, so there is an incentive for industries to support regulation. When an industry acquires regulation, however, it can be expected that the benefit to the industry usually is less than the damage to the rest of the community (Stigler, 1971). Empirical studies of public utility industries suggest that although price regulation generally constrains prices below the level which an unconstrained monopolist with a legally exclusive franchise would choose, they may not be lower than those which would exist under a fundamentally different, more competitive, industry structure (Joskow and Rose, 1989).

A second group of interests is made up of bureaucrats, including both those concerned with the regulation of the activities of private firms and those who manage direct public interventions. Rent-seeking by bureaucrats includes the pursuit of self-interest so as to increase their areas of power and responsibility, to maximise the resources at their disposal, and to inflate costs. These are activities to which they might be even more prone than politicians, given that bureaucrats usually:

1. do not have a clearly defined set of objectives against which to measure their performance because of conflict among policy objectives and because an operational definition of welfare is lacking;
2. have considerable scope for discretion;
3. in the case of regulation have ample scope to maximise their own interests and pursue private goals since offences are usually poorly defined. Moreover, it is usually hard to create incentives for effectiveness in many public agencies and the failure of a regulatory agency to function effectively can undermine the potential benefits from competition and private sector participation.

The incentives to rent-seeking by bureaucrats include both direct and indirect rewards. The direct reward is the maximisation of lifetime income. The indirect rewards include the consideration, on the one hand, that most bureaucrats find their jobs interesting and intellectually satisfying, while on the other, some may have their own economic theories which they are likely to be tempted to apply in their work. The great public water management institutions such as the Bureau of Reclamation and the Corps of Engineers in the United States, the *Secretaría de Recursos Hidráulicos* in Mexico or the Scientific Research Institute on the Irrigation Problems of Central Asia (SANIIRI) of the USSR were always known for their technical expertise and high morale.

Vulnerability to capture and opportunity for rent-seeking increase for regulators, in particular, where the regulatory regime provides them with high levels of discretion and entrusts them with open-ended powers while leaving their objectives ill-specified and their duties vaguely defined (Helm, 1993).

Other groups of rent-seekers can be readily identified. For example, workers employed in the water sector can be expected to capture a share from the possible monopoly rent either in the form of wages and other benefits, such as security of employment, or by not having to work too hard. Workers commonly oppose privatisation. A further group is made up of influential consumers. These can use their influence with politicians to achieve favourable tariffs, which helps explain, at least in part, the prevalence of uniform tariffs and cross-subsidisation for drinking water supply and sanitation. They may also be able to determine the composition of the investment programme. Empirical studies suggest that in public utility industries, the structure of prices across classes of customers often reflects income distribution and political objectives, rather than efficiency considerations (Joskow and Rose, 1989).

Special-interest lobbies, which demand services which are financially unfeasible and which will require either subsidy from general public-sector revenues or a cross-subsidy from potentially profitable services, form another rent-seeking group. The major costs of rent-seeking in this last case arise from the need to use inefficient methods to transfer funds to the potential beneficiary, because efficient methods, under these circumstances, would be too open to public scrutiny (Tullock, 1987).

Capturing the Public Interest

Government failure or capture of the public interest occurs when an agency or regulator falls under the undue influence of one party, either the government, the regulated firm, its competitor or rival, consumers, or some other interest group. It happens because public decisions affect everyone's welfare.

> When government intervention, rather than actual market performance, determines which firms are winners and which are losers, corporate executives have an incentive to devote resources to lawyers and consultants rather than to scientists and engineers. The hearing room rather than the industrial laboratory becomes the focus of attention (Berg and Foreman, 1995).

There are two broad kinds of institutional and regulatory capture: that which occurs in the legislature as policies are formulated and legislation passed, and that which occurs as decisions are taken once the regulatory framework is in place. Members of interest groups, including the regulated firms, are organised and have strong incentives to exercise political pressure on legislators and bureaucrats in order to affect decisions. As a result, these compact, well-organised groups will tend to benefit more from public decisions than broad, diffuse groups (Peltzman, 1989). As interest groups must compete with each other for influence, government intervention will tend to preserve a politically optimal distribution of rents among them.

One obvious response to the possibility of capture is to reduce any interest group's stake in the public decision-making process. However, if interest groups also make a positive contribution to the process, for example, bring new information, it may be socially desirable to increase their stakes so as to induce them to acquire information (Laffont and Tirole, 1990).

There is also always asymmetry of information between the regulators and the political authorities and between the latter and voters. All bureaucracies are characterised by the attenuation of control. Lower-ranking officials keep and control most of the information needed to assess their performance. Much of what they do is, therefore, unknown to those of higher rank (Tullock, 1987). The higher the official, the less he or she knows of what occurs at the lower level of the bureaucracy. Consequently, high-ranking authorities usually receive only an 'official' view of activities. They have neither the information nor the expertise and resources to evaluate these reports against vaguely defined or unstated objects. Voters are never well-informed about their votes because the effect of their own vote on their well-being is small. The information at a voter's disposal tends to be biased in the direction of his or her own special interests. Only in the absence of such informational asymmetries would voters and political authorities be able to effectively control their agents. Only then would bureaucrats be unable to implement policies favouring special interest groups at the expense of society as a whole (Laffont and Tirole, 1991).

Given that, in many countries, there has been a long history of state interference in the provision of water-related services, the possibilities for establishing non-political management systems seems to be fraught with many obstacles. For example, in many countries, the tariffs of water-based services have traditionally been unrealistically low and politically controlled. If they were to rise to reflect real economic costs, there would likely be some political pressure to minimise any increases. In part, privatisation can aid in resolving this problem because it increases the transaction costs of government interference in the workings of the firms. The efficiencies that privatisation can achieve stem fundamentally from the insulation it provides from inefficient political and self-serving influences (Willig, 1993).

The administrative capacities of many governments are already strained by the weight of existing activities. Government intervention in water management is not free of cost and even efficient regulation is a complex and expensive task. The managerial and financial resources needed are scarce, particularly in the public sector. Given these considerations, a too elaborate and complex public management system not supported by adequate capacity and commitment and operating under budgetary constraints has been shown to produce efficiency losses greater than those it is intended to avoid.

This should not mean that government intervention in water management should never be contemplated. It does mean, however, that in considering the establishment of water management and regulatory systems governments should be open-minded in judging the various alternatives that are available and be cautious in developing too elaborate systems in environments without traditions of strong public service. It also underlines the need to target intervention on the areas where market failures are most pronounced, to pay attention to the costs and benefits, and to design regulatory mechanisms to maximise the benefits in relation to the costs. Neither privatisation nor user-management regimes can, of themselves, release governments from their responsibility to provide their populations with reasonable and equable access to basic water-related services. This responsibility has to remain with government and in the area of public policy.

REFERENCES

Berg, Sandford V. and R. Dean Foreman (1995), 'Price cap policies in the transition from monopoly to competitive markets', *Industrial and Corporate Change*, **4** (4), 671–681.

Crane, Lyle E. (1969), *Water Management Innovations in England*, Washington: Resources for the Future.

Day, J.C., Enzo Fano, Terence R. Lee, Frank Quinn and W.R. Derrick Sewell (1986), 'River basin development' in Robert W. Kates and Ian Burton (eds), *Geography, Resources, and Environment, Volume II, Themes from the Work of Gilbert F. White*, Chicago: University of Chicago Press, pp. 116–152.

Fair, Gordon M. (1961), 'Pollution abatement in the Ruhr district', in Henry Jarrett (ed), *Comparisons in Resource Management*, Baltimore: Johns Hopkins Press, pp. 142–171.

Frederick, Kenneth D. (1996), 'Water as a source of international conflict', *Resources*, **123**, 9–12.

Frederiksen, Harald D. (1992), *Water Resources Institutions*, World Bank Technical Paper No. 191, Washington D.C.: World Bank.

Germany, Federal Ministry of the Environment (1998), *Environmental Policy in Germany*, Bonn: Federal Ministry of the Environment.

Helm, Dieter (1993), 'The assessment: reforming environmental regulation in the UK', *Oxford Review of Economic Policy*, **9** (4), 1–13.

Holtram, Gerald and John Kay (1994), 'The assessment: institutions of policy', *Oxford Review of Economic Policy*, 10 (3), 1–16.

Information Highway to the Global Environment (IHGE) (1995), A Survey of *Environmental Monitoring and Information Management Programmes of International Organizations*, Edition 3.4, Internet.

Joskow, Paul L. and Nancy L. Rose (1989), 'The effects of economic regulation', in Richard Schmalensee and Robert D. Willig (eds), *Handbook of Industrial Organization*, Vol. II, North-Holland, Amsterdam: Elsevier Science Publishers, pp. 1449–1506.

Kühner, Jochen and Blair T. Bower (1981), 'Water quality management in the Ruhr area of the Federal Republic of Germany, with special emphasis on charge systems', in Blair T. Bower, Rémi Barré, Jochen Kühner and Clifford S. Russell (eds), *Incentives in Water Management France and the Ruhr Area*, Research Paper R-24, Washington: Resources for the Future, pp. 211–302.

Laffont, Jean-Jacques and Jean Tirole (1990), Accounting and collusion, mimeo, M.I.T., as quoted in Laf and Tirole (1991).

Laffont, Jean-Jacques and Jean Tirole (1991), 'The politics of government decision making: regulatory institutions', *The Quarterly Journal of Economics*, **106** (4), 1089–1127.

Lambert, David K., Dimo Dichev and Kambiz Raffiee (1993), 'Ownership and sources of inefficiency in the provision of water services', *Water Resources Research*, **29** (6), 1573–1578.

Lee, Terence R. and Andrei Jouravlev (1997), *Private Participation in the Provision of Water Services*, Serie Medio Ambiente y Desarrollo, No 2, Santiago, Chile: Economic Commission for Latin America and the Caribbean.

Lovei, Laszlo and Dale Whittington (1991), *Rent Seeking in Water Supply*, Report INU 85, The World Bank, Sector Policy and Research, Infrastructure and Urban Development Department, Discussion Paper, Washington D.C.: The World Bank.

Maass, Arthur, Maynard Hufschmidt, Robert Dorfman, Harold Thomas Jr., Stephen A. Marglin and Gordon Maskew Fair (1961), *Design of Water Resource Systems*, Cambridge, Massachusetts: Harvard University Press.

Peltzman, Sam (1989), *The Economic Theory of Regulation after a Decade of Deregulation*, Brookings Papers on Economic Activity, Microeconomics, Washington, D.C. : Brookings Institution.

Phillips, Charles F. (1993), *The Regulation of Public Utilities. Theory and Practice*, Arlington, VA: Public Utilities Reports Inc.

Schramm, Gunter (1980), 'Integrated river basin planning in a holistic universe', *Natural Resources Journal*, **20** (4), 787–806.

Stigler, George J. (1971), 'The theory of economic regulation' *The Bell Journal of Economics and Management Science*, **2** (1), 3–21.

Teclaff, Ludwik A. (1967), *The River Basin in History and Law*, The Hague: Martinus Nijhoff.

Tullock, Gordon (1987), 'Rent seeking', in John Eatwell, Murray Milgate and Peter Newman (eds), *The New Palgrave. A Dictionary of Economics*, vol. 4, London: The Macmillan Press Limited.

Werner, Frank-Uwe (1997), *Die Bedeutung von Wassermärkten für ein nachhaltiges Wassermanagement in ariden Regionen*, Diplomarbeit am Institut für Geographie, Fakultät für Geowissenschaften, Heidelberg: Ruprecht-Karls-Universität.

White, Gilbert F. (1957), 'A perspective on river basin development', *Law and Contemporary Problems*, **22** (2), reprinted in Robert W. Kates and Ian Burton (eds) (1986), *Geography, Resources, and the Environment. Volume I, Selected Writings of Gilbert F. White*, Chicago: University of Chicago Press, pp. 39–79.

White, Gilbert F. (1963), 'Contributions of geographical analysis to river basin development', *Geographical Journal*, **129** (4), 412–436.

White, Gilbert F. (1969), *Strategies of American Water Management*, Ann Arbor, Michigan: University of Michigan Press.

Willig, Robert D. (1993), 'Public versus regulated private enterprise', in *Proceedings of the World Bank Annual Conference on Development Economics, 1993*, Washington, D.C.: The World Bank, 15–172.

Wirl, Franz (1992), 'The European power industry – characteristics and scope for deregulation', *OPEC Review*, **16** (2), 137–156.

World Bank (1993), *Water Resources Management*, A World Bank Policy Paper, Washington: World Bank.

World Resources Institute (1992), *World Resources, 1992–93*, New York: Oxford University Press.

3. Allocating Water among Competing Users

The most serious issue, among the many matters that water management has to consider, is that of the allocation of water among competing uses and users. The issue of allocation overshadows all other aspects of water management, including the difficulties of managing water quality, controlling flows and all the remaining myriad questions involved in managing water. Unfortunately, given its significance almost everywhere, the methods used for water allocation are woefully inadequate to achieve an effective, efficient and timely distribution of water among uses and users. This failure to allocate water efficiently is the root cause of the evermore widespread perception that water is becoming scarce to the extent that in many places we are facing a crisis. The result of the existing degree of inefficiency in water allocation is that it is impossible to know at any place or time whether the use of water is efficient or not.

In most places, water is allocated either according to some inherited historical decision or by reason of bureaucratic processes subject to political pressures. As a result of these systems for allocating water, if they can be called such, changing the distribution of water to respond to changing water demands is at best slow and inefficient or, only too often, not possible at all. For example, under riparian systems of water rights, unless a potential new user can purchase riparian land the chances of obtaining access to a source of surface water can be very slim. Even when licensing or permit systems which consider the possibility of reallocation are used, any request for a change in the existing distribution of permits often must pass through a complex bureaucratic and political system for approval.

The largest user of water is agriculture, even in countries where irrigation is not a common practice (Figure 3.1). Only in economies dominated by pastoralists and smaller island countries does agricultural use not tend to dominate. Reallocation of water to improve the overall efficiency of water use, in most cases, means, therefore, transferring water from agriculture to

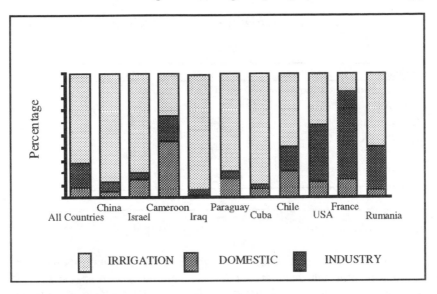

Source: Seckler et al., (1999).

Figure 3.1 Water use by sector, selected countries

other sectors. Generally speaking, the use of water in agriculture, particularly in irrigation, is inefficient, partly because farmers receive no incentive to be efficient and further, because governments commonly subsidise inefficient agricultural production thus compounding the inefficient use of water. It should be stated that the fact that the use of water in agriculture is generally inefficient does not mean that water use by other sectors is anywhere near the optimum. There are technological reasons, however, which result in other sectors achieving a higher value-added per unit of water used than in agriculture, even in the absence of economic incentives. Unless, however, water is treated as an economic good and appropriately priced, it is really not possible to measure the economic efficiency of its use by any sector.

A separate question from the allocation of water among uses and sectors of use is whether the total amount of water used is at all close to the optimum. Under existing water management practices, where allocation is largely arbitrary, it is most unlikely that water use would be close to the optimum. It is most likely that overall water use, in the absence of the signals provided by prices, will exceed the optimum. What we do know is that, in practice, the use of water for any particular purpose, whether residential, industrial or agricultural, shows variations of several orders of magnitude among uses with similar characteristics (Table 3.1).

Table 3.1 Variance in industrial water withdrawals (gallons per unit of production)

Industry	Maximum	Minimum
Thermal power station (kWh)	170	1.3
Oil refining (gallon of crude oil)	44.5	1.7
Steel (ton of finished steel)	65,000	1,400
Soap and edible oils (lb.)	7.5	1.6

Source: Anderson (1985).

Long before the recent concern with water scarcity, economists interested in water management have argued for the need to recognise water as an economic good. They have insisted that water cannot continue to be treated as having a 'unique importance', but should take its place simply as one good, one natural resource commodity, among all others:

> This is not to deny that, as a commodity, water has its special features: for example, its supply is provided by nature partly as a store and partly as a flow, and it is available without cost in some locations but rather expensive to transport to others. Whatever reason we cite, however, the alleged unique importance of water disappears upon analysis (Hirshliefer, De Haven and Milliman, 1960).

If water is treated by water managers as the economic good that it is, then it is possible to govern its allocation through the market. For many years, it has been widely recognised in the economic literature that, as we have already alluded, in the absence of markets it is difficult, if not impossible, to evaluate the real demand for water-related services because demand functions cannot be estimated (Fox and Herfindahl, 1964). In the place of markets and the signals for efficiency in investment decision-making which they provide numerous, elaborate and unsatisfactory substitutes have been suggested and applied. All these substitutes have in common that they provide only poor, if not incorrect signals, that they are essentially arbitrary and that they provide no real solution to the problem of achieving an efficient allocation of water. The only solution is to place as great a reliance as possible on prices and, therefore, on markets in the process of the allocation of water and the related investments in hydroelectric power generation, irrigation schemes and all the other productive uses of water. If efficiency is the goal, and if markets are the most efficient way to allocate scarce resources, then the role of administrative allocation should obviously be restricted to those few areas, if any, where markets cannot be developed and to the regulation of natural monopolies.

It must be carefully considered, however, when advocating the introduction, extension or maintenance of any public intervention in water allocation, how effective it can be. As was commented in the debates almost 30 years ago on this issue in the United States:

> that while public intervention was necessary, it need not be sufficient for improvement in efficiency. For intervention to be also a sufficient condition for improvement in efficiency, appropriate criteria must be developed and, assuming in the final analysis that there is a feasible way to do so, applied with sufficient fidelity to ensure that the objectives of public intervention in the interests of efficiency are reasonably approximated (Krutilla, 1966).

In discussing both public policy towards water management and the issue of using prices and introducing markets as a major tool for water, it has to be understood that 'controlling the use of water courses is a basic economic problem of resource allocation' (Freeman and Haveman, 1971). Facts equally true for both the quality and quantity aspects of allocation.

To come closer to the present and to current water management issues, the increasing private participation in water management has brought with it as a corollary the wider opening of water management to market forces. It has also increased the interest in directly employing prices and markets as the main tools for the allocation of water among different uses. One sign of this interest is the amount of literature discussing the experience of the few places where water markets exist, particularly the amount of attention given to the Chilean experience.

In Chile, water markets were adopted as the principal tool for water allocation under the 1981 revision of the water code. The reforms have placed an almost total reliance on market signals in the allocation of water, free from bureaucratic intervention:

> In the 1980s, Chile shifted from a system where water rights were State owned, to a system of private enterprise-market oriented rights. ... the policy helped users to secure water rights, both tradable and transferable. ... In order to achieve higher efficiency, the policy allows for market allocation of water within and between sectors (Gazmuri, 1994).

The adoption of a market approach to water allocation in Chile has attracted considerable attention from all over the world, although serious economic analysis remains largely absent from the discussion of the Chilean experience. The findings of the one economic analysis that has been made suggests the importance of remedying this lack of serious economic evaluation (Hearne and Easter, 1995), but there is certainly much more literature on the current Chilean system than on any previous water allocation system adopted in any other country. The Chilean experience will be discussed in detail later in this chapter.

Here the discussion will centre on the means to incorporate the use of market signals through prices into water management with the objective of improving efficiency in the allocation of water. The experience in the use of prices and markets in water allocation in the two places with significant application, the western United States and Chile will also be discussed. The focus will be on the issue of the quantitative allocation of water.

Once markets are introduced and property rights established in water, it can be expected that transfers of water rights will occur whenever the net benefits from a reallocation are positive, until marginal values, after subtracting transaction and conveyance costs, are equalised among water users, uses and locations. Trade will continue until all water users are satisfied, in economic terms, become indifferent about buying or selling water rights. The difference in the value of water in alternative uses and locations precipitates market transactions. This difference must be large enough to outweigh the transaction and transportation costs of obtaining water through the market. Water markets are unlikely to emerge, be active or operate effectively where water is in surplus. If there is more than sufficient water to satisfy all needs, the opportunity costs of water will be zero and there will be no incentive to trade. It has been observed that it is for this reason that market activity intensifies in periods of insufficient water supply and becomes less active or latent under normal supply conditions.

In Chile, for example, it has been noted that water markets are more active both in those basins in the north of the country where water is more scarce and in periods of drought (Donoso, 1998). The relative inactivity of the water market in many parts of the country has been attributed in part to the fact that potential buyers have other and cheaper options for obtaining water. For example, in central and southern Chile, groundwater supplies are still relatively plentiful and little exploited. Other factors that have limited transactions in water rights include possibilities for improving efficiency in the use of water and the storage on the farm of water from winter flows. To cite another example, in Texas, transfers of surface water rights have been theoretically possible for several decades. They did not begin, however, until the first river basin was fully adjudicated, that is, all the water was allocated. Major trading activity has been concentrated in the Lower Rio Grande Valley where there is little groundwater (Chang and Griffin, 1992).

INTRODUCING WATER MARKETS

The decision to introduce a system of tradable water rights requires the consideration of many issues so that the water market can function smoothly and equitably. Among the important issues that must be considered is the initial allocation of rights. In the initial allocation it is necessary to take into account the acquired rights of existing water users, but at the same time, it may be necessary politically to limit the windfall gains this may produce. The establishment of individual rights must not be seen to come at the expense of society as a whole. It is also essential to ensure that rights are clearly and securely defined and appropriately registered. Whatever decisions are taken and whatever policies are adopted, the system must be as simple as possible. Rights and obligations must be clearly established in law and, if the market is to rule, the intervention of government must be kept to a minimum.

The Initial Allocation of Water Rights

The introduction of a market with tradable water rights requires the prior determination of the total number of water rights to be allocated to existing water users. The approach taken in this initial allocation can have a substantial impact upon the implementation and efficiency of the subsequent use of water marketing as the main means for water allocation. How rights are allocated is crucial to the acceptance or rejection of a water market by existing users.

Theoretically, as long as there are markets where the rights are freely transferable the final equilibrium allocation of water rights will be the same, regardless of their initial distribution. However, this can only be the case in a setting with many buyers and sellers, full information, zero transaction costs and other stringent assumptions. Given these assumptions, at least in theory, this 'implies that under the right conditions the initial allocation can be used to pursue distributional goals without interfering with cost-effectiveness' (Tietenberg, 1995).

Where information, bargaining, contracting and enforcement are not without cost, and where there is market power or resource immobility, the initial distribution of water rights can and does affect the efficiency with which a water market will reach equilibrium. The initial distribution of water rights can affect the quantity of transactions, the equilibrium allocation of rights and the aggregate benefits of water marketing. If transaction costs are sufficiently large they can preclude trade with the result that the initial resource allocation will be retained. There is, therefore, a trade-off between promoting efficiency and equity in any initial distribution of water rights.

Alternative initial assignments of water rights among users will result in different sets of bargaining relationships and different patterns of water use and transfers. Future outcomes, therefore, will depend on the initial distribution of rights but it has been accepted that one original distribution of rights will give rise to one set of market outcomes and a different distribution of endowments will give rise to a different set of outcomes. It is possible for both sets of outcomes to be considered Pareto efficient if resources are divided originally in an equitable manner (Chan, 1995). Consequently, future economic efficiency will depend on market competition and not on the relative merits of the original distribution of rights.

A market will reallocate water rights through voluntary transfers between buyers and sellers. The future impacts of market transactions, however, on income distribution – how the scarcity rents from water are distributed, who has the protection of the State to use water as they wish, who must pay to obtain water rights, and who receives payments – will depend on the initial allocation of rights. The initial choice is political and the choice must be made with explicit consideration of both efficiency and equity.

There is a need in introducing a market system to draw attention to the distribution and recognition of existing rights. This can be done through campaigns of public information, as well as by offering legal and technical advice and by providing assistance to disadvantaged groups. In Chile, for example, the government has adopted and pursued for many years a programme to facilitate the legalisation of the property titles to water rights (Ríos and Quiroz, 1995).

A further issue that must be addressed is whether a government should try to recover the capital costs of historic public investment in water-related infrastructure. Various solutions have been proposed ranging from outright prohibitions of transfers and other restrictions on trading for the beneficiaries of such schemes, to ignoring the problem and allowing transfers without any restrictions. The first policy alternative, if adopted, is likely to result only in stifling incentives to transfer and conserve water resources. The second alternative will facilitate the reallocation of water to higher value uses, but at the expense of windfall gains to existing users. Intermediate positions include the imposition of a windfall tax, dividing the windfall according to a formula which encourages transfers and yet permits the government to share in it (Gould, 1989). They also include greater reliance on alternative mechanisms for initial allocation, such as auctions. It is clearly better, however, to err on the side of fewer restrictions and not to prevent market transfers. It is perhaps best to rely on taxes to restrict unfair individual gains.

Other issues to be considered in the initial allocation of water rights include such questions as establishing minimum flows to protect aquatic habitats and other uses that cannot compete in the market for water. Where an initial allocation of water for minimum flow maintenance is not possible because historic uses have pre-empted the total supply, an argument can be made for a one time reallocation of water as a means of redressing past policy deficiencies which resulted in over-allocation (Griffin and Boadu, 1992). These questions, however, can be considered later. For example, in Chile, although no provision for the maintenance of minimum flows was included in the law, new water rights can be granted only if this does not affect the rights of third parties. In recent years, the interpretation of the rights of third parties has been expanded to include environmental protection and the maintenance of minimum ecological flows (Peña, 1996a).

A similar argument can be made to take advantage of the introduction of a new system of water allocation to reapportion water in favour of previously disadvantaged groups in the population. If such a decision is made, it is particularly important to assure other water users that such interventions to reallocate rights will not recur. A better alternative, taking advantage of the existence of the market, might be for the government to acquire water rights at the going price for any desired social purpose.

Alternative procedures for allocating rights
Where a water market is introduced to replace administrative allocation, a government may choose to distribute water rights based on some regulatory distribution rule, usually the historic record of possession of permits for water use under the previous system. This is sometimes called 'grandfathering'. Where such records are absent, other benchmarks, such as land holdings, might be used. The alternative to such methods would be for the government to sell water rights through auctions. Auctions should be used so as to ensure that the sale of rights would both be equitable and match supply with demand. Obviously, any combination could be used, such as allocating a proportion free of charge based on historic use and the rest by auction. Leasing is a possibility, but it is likely to be less attractive as it would provide a lower level of security and probably reduce future investment in water-related activities.

Basing the initial allocation of water rights on historic water use is easiest. This is the system, therefore, that has usually been used. Governments could achieve greater efficiency gains through auctions, but there are potentially high political costs if many existing users fail to obtain rights. Grandfathering represents effectively a transfer of wealth to existing water users. Auctions would ensure that the wealth represented by water rights is transferred to the society as a whole and windfalls are avoided. Existing users prefer grandfathering to auctions because it preserves the status quo. It can be argued, however, that because the ownership and ability to transfer water rights provides a new asset, it is reasonable for the government to be able to sell at least some of the rights.

Selection of the procedure for initial allocation raises a number of other equity questions:

1. whether equity considerations should apply only to existing water users, or to all citizens;
2. whether existing water users should have to pay for water rights at all;
3. whether and how environmental and other instream interests can enter the allocation procedures (Lyon, 1982).

With auctions, water rights will tend to go to those to whom they are most valuable; hence an auction immediately promotes efficiency. Under grandfathering, increases in efficiency must await subsequent sales and purchases of rights in the market.

Grandfathering can provide ample opportunity for administrative discretion unless the allocation is based upon reliable information that is difficult to manipulate, for example the records of the users, themselves. It can invite lobbying efforts to protect vested interests and maintain the status quo and may give perverse incentives to existing rights holders. Where historic records are weak, anticipation of grandfathering could encourage users to expand their withdrawals in order to qualify for more water rights.

If regulators do not act to prevent such strategic behaviour, the allocation process can reward the least efficient users. In contrast, auctions not only prevent such negative consequences and mitigate the appearance of favouritism or secret negotiations; they also remove the incentive for existing water users and other interest groups to engage in rent-seeking activities, which waste resources. Auctions also readily identify appropriate prices and make the value of water more explicit when introducing a market.

The main advantage of allocating rights on the basis of historical use, and the reasons for which this is the usual alternative selected, is that it avoids conflicts. It also reduces the opposition of existing water users to the introduction of a market. Farmers usually argue, reasonably, that they are entitled to receive water rights without charge because they have already paid for the rights implicitly in the purchase price of the land. With the assignment of rights based on historical use, there is no new financial burden created by the need to pay for water rights.

National Experiences in Introducing Markets

In practice, most countries have recognised existing water uses at the time of introducing market assignment of property rights to water, both to protect existing uses and to prevent opposition. Where there is a well-functioning traditional registry of water rights and where all water rights are honoured in the water distribution system, it is usually sufficient to recognise existing rights. This is so whether or not water users reregister them in a property rights register. If there is no traditional registry or if the volume of water rights exceeds available supplies, complications can arise since other solutions such as need, land area, or other benchmarks may not serve.

The initial allocation of water rights in Chile, at the time of the promulgation of the 1981 Water Code, which reintroduced and amplified private property rights in water, was based on the historical allocation of water. These rights could be overridden, however, in favour of those who had been making 'effective' use of the right for the five years prior to the promulgation of the law. The availability of relatively good records held by user associations facilitated the allocation process and made it possible to honour the historic allocation (Cestti and Kemper, 1995). Water rights have always been assigned to the applicant without charge and without any obligation that the water be put to use.

Other countries have also adopted the historical allocation approach. In Mexico, under the reforms introduced in 1992, water rights were allocated on the basis of the formal and informal water rights already held (Holden and Thobani, 1995).

There are, however, examples of the use of auctions for the allocation of water rights, usually for water not allocated in an initial allocation process or made available because of new infrastructure. An auction was used for the allocation of new water supplies made available from a new reservoir in the State of Victoria, Australia. The completion of the dam made 35,000 million litres of water available to be allocated in the form of 15-year diversion licences. A number of methods were considered for allocation. The initial decision to require sealed bids was discarded because of strong resistance from farmers in favour of a more open auction process.

The following rules were adopted:

1. Participants had to indicate the maximum volume of water that they were interested in acquiring and its intended use.
2. Participation was limited to private irrigators with legal access to the river from which the water was available and to landowners or lessees in the basin. Users in public irrigation districts and urban areas, and speculators were not eligible to participate in the auctions.
3. A reserve price was established but not disclosed at the auctions.
4. Purchases by a single land holding were limited to 10 per cent of the volume being offered, although anyone with multiple land holdings could purchase more than the limit.
5. The sale of water was staged; a minimum purchase set for each stage. Bidders were required to make at least the minimum purchase corresponding to the stage in which they participated. They could participate in several or all stages, but had to register to make at least the minimum purchase in a particular stage.
6. Bidders competed based on their willingness to pay for 1 million litres. The highest bidder could purchase the volume of water desired at this price. Other bidders were allowed to purchase any remaining water at the same price. If they desired more water than that available, bidding was reopened.

Of the 31,000 million litres offered for sale, 23,000 million were sold through auction. Prices declined with the bulk of the water purchased at the reserve price or slightly higher. Lower prices were paid for larger volumes of water, and higher prices for smaller volumes, usually bought for the production of high value crops. A quarter of the 200 successful bidders were new irrigators. All the water offered in the first two auctions was sold, but not in the later auctions. Consequently, eligibility was expanded to include users from other basins, in an effort to encourage competition and to broaden the market, but water remained unsold. Demand and prices declined as more auctions took place. The reserve price became known before the later auctions were held and bidders became unwilling to bid higher.

The auction in Victoria was a success, but the restrictions, although they improved equity, did so at the expense of introducing a degree of inefficiency. Competition was constrained. The chances of transferring greater volumes to higher value uses were reduced. The benefits for public finance were limited, and it was not possible to extract all of the gains from trade (Simon and Anderson, 1990; Cestti and Kemper, 1995).

THE DESIRABLE ATTRIBUTES OF A MARKET SYSTEM FOR WATER ALLOCATION

In support of the exclusion of water from the market, it has traditionally been argued that water has not only special physical and economic characteristics, but that it possesses a cultural importance that sets its off from all other resources used by man. Such claims, as powerful as they have been in determining that water should not be treated as an economic commodity, are difficult to maintain, however, when any rational comparison is made between water and other natural resources. There is nothing unique to water except its innate mobility. This does pose an added complication to the establishment, definition and enforcement of property rights, which are the essential foundation of any market allocation mechanism. These characteristics of water are not such as to 'rule out either the possibility or the desirability of using prices and regulated markets to introduce economic incentives to restrain use, encourage conservation, and facilitate reallocation of supplies' (Frederick and Kneese, 1988).

The force of the idea that water is different results in the fact that most of those arguing for the treatment of water as an economic commodity and of using the market to achieve the objective of efficient allocation, emphasize that they mean a regulated market. But, do unregulated markets exist for any commodity in contemporary societies? Would anyone propose relying on an entirely unregulated market structure for water allocation? Surely, it is generally accepted that there is a requirement for any market to be subject to regulation.

It must be recognised in this discussion that a water market is a water management tool. It is, however, a tool that spreads the burden and difficulties of water management among a larger population, permits greater participation in management decisions and can introduce greater flexibility into management systems. The establishment of a water market demands new skills and new attitudes from the public administration, judicial systems and water users, as well as investment in the registration of rights, monitoring and measurement systems and, possibly, in improving water distribution and transportation systems. On the whole, however, it is the case that '... the prerequisites needed for a viable water market are the same as those needed for good water management' (Simpson, 1994).

In a water market, water is allocated at a price set by the free exchange of a property right to the use of water either for a limited period of time through a lease or in perpetuity by a sale. The water market is the institution, formal or informal, which facilitates the exchange of water rights among buyers and sellers. It is the interactions between the buyers and sellers of rights that comprise a water market. Water markets can be distinguished from other processes for water allocation by:

1. The motivating force is the voluntary perception by both buyers and sellers that the transaction is in their own best interest given the alternative opportunities available to them.
2. No central authority determines price and other terms of transfer, although it may condition or regulate them.
3. The price is generated through voluntary transactions negotiated between willing buyers and sellers.
4. The transfer of water is the real purpose of the transaction and the value of water is established independently of the value of other goods and services involved in the transaction. A water market only exists where water rights are commodities with an identity distinct from other real property.

For a market to work, it is necessary only that there be a tradable margin, even if it is only a small part of the total supply. Moreover, due to the nature of water use, water markets can normally be expected to be relatively small or 'thin'. The number of transactions will be a function of many factors, one of the most important of which will be the historical allocation of water among users. The number of transactions does not say much, if anything, about the ability of a market to efficiently reallocate resources in response to changes in their marginal productivity.

The desirable characteristics for a water allocation system include both flexibility and security (Figure 3.2). At the same time, all costs and benefits should be reflected in the decisions that participants take. If water users do not face all the costs and benefits associated with their decisions, although their decisions may be beneficial to them, they could be inefficient from an overall social perspective. The system must also be predictable, and show equability and fairness, with the capacity to reflect collective, public and social values (Howe, Schurmeier and Shaw, 1986b). Therefore, the most important characteristics are flexibility and security, together with the requirement that, as far as is feasible, buyers and sellers bear all the costs associated with their decisions to undertake a market transaction.

Economic growth and water use efficiency requires the achievement of a balance in the allocation of water between flexibility and security. Although

Figure 3.2 'Flexible, secure and simple', an advertisement, from a Santiago, Chile newspaper, offering a water right for sale

the degree of security of tenure of the right to use water can cause a reduction of flexibility, and greater flexibility in the system reduces security, both can be achieved simultaneously as long as users can voluntarily respond to incentives for reallocating water supplies. It can be strongly argued that, as in the rest of the economy, the use of the market as a system for water allocation more often meets the requirements for an effective and fair allocation system than any other possible mechanism.

The Flexibility of Water Markets

Water markets are flexible because markets are by their very nature a decentralised and incentive-oriented institution, rather than centralised and regulatory. Transferability of water rights in the market provides the freedom to reallocate water as economic, social and environmental demands and conditions change.

> In a dynamic society with continually changing values, it is this transferability which insures flexibility. Entrepreneurs continually have new and better ideas of how to utilise resources. It is their offers to buy and sell these resources that generate progress. If transferability is not allowed, there is no effective way for the system to respond to changes in demand and supply (Anderson, 1985).

The rationale of a water market is exactly this granting to water users of flexibility in their use and transfer decisions. Once all water is allocated, any user can establish or expand activity only by purchasing water rights from other users.

With a market for freely transferable water rights, by definition, the marginal values for water, less transaction and conveyance costs, are equated across water users, uses, and locations. The equalisation occurs because the market provides strong incentives for water users to reallocate water rights whenever reallocation would generate positive net benefits. The transferability of water rights in the market enables new uses and users to emerge and obtain water supplies at a reasonable cost. It also prevents waste and encourages water conservation. It provides a continuous incentive for adoption, research and development of superior water conserving and production technologies. A market-based system of water allocation will be, therefore, both resilient to shocks and open to taking advantage of opportunities.

Security of Tenure

Markets require security of tenure, which in turn helps encourage efficient use, resource conservation, and capital investment. Security of tenure of water can also help strengthen and consolidate the autonomy of water users-organisations. Security of tenure and the possibility to acquire water rights in the market encourage investment and growth in activities that require secure water supplies. The fact that in a market water rights are reallocated by voluntary exchanges allows market systems to defuse potential political conflicts over water allocation.

The political dimension of the implementation of a system of secure and transferable water rights arises from the fact that the definition and clarification of property rights 'offers significant potential for minimising the costs of conflict of use in multiple use resources' (Pearce, 1989). Market transfers are always voluntary transactions in which traders will only participate if they believe that it is in their best interest given the alternative opportunities available to them. 'Market systems have a tendency to defuse political conflict, largely because anyone who obtains a resource must pay the prior owner a price that satisfies that owner' (Williams, 1983). Administrative allocation, in contrast, often generates intense conflicts because granting a water right to one user necessarily precludes another and there is no automatic pecuniary compensation for the basin of origin. Water markets change the nature of bargaining over water transfers: 'instead of political wrestling, with the losing region defeated by the winning region, the bargaining can become a process of mutually advantageous exchange' (Williams, 1983).

Markets can also reduce conflicts among environmental interests, water suppliers, and polluters through providing incentives for water conservation and wastewater treatment. Ecological economists often prefer property-right systems to pricing systems because 'property-right systems define the ecological limits and then leave the market to work out what prices and charges are necessary to keep use within those limits across space and through time. On this view, property-right systems tend to be ecologically more dependable than pricing systems ... governments routinely fail to vary prices in response to changing economic conditions and opportunities ... When a property-right is used to define the limit, however, market processes take over' (Young, 1997).

External Effects

While market transactions require a guarantee of security of tenure to buyers and sellers, the rights of third parties can be vulnerable to externalities from water transfers. Thus, to ensure that market transfers do indeed produce net social benefits, water marketing must be conducted in an institutional framework that causes the buyers and the sellers to take account of third party impacts.

By setting a market-clearing price and making current and potential market participants aware of the ability to sell and buy at that price, if desired, every market directly confronts water users with the real opportunity cost of their use and transfer decisions and forces them to take this opportunity cost into account. Water users, facing the full opportunity costs of their decisions, including the impact on third parties, will take only those actions that generate positive net benefits both for themselves and for society as a whole. If water users are to bear the full opportunity costs of their actions, the market must be fairly competitive and water rights freely transferable. This means that public policies should seek to facilitate market operation. Policies should also see that competition is not undermined and that uncompensated costs are not imposed on third parties, and ensure the transferability of water rights over as wide a geographical area and among as wide a variety of participants as possible.

In theory, a water agency could fix prices at levels approaching the opportunity cost of water, but it would have to be endowed with uncommon wisdom, efficiency and foresight. The principal advantage of a market is the ability to gather, process, and use information. Demand and supply conditions continuously change, and this information is fragmented and dispersed among all actual and potential water users and is both time- and place-specific with a high variance across localised ecosystems. If public authorities could possibly have the immediate information necessary to make trade-offs among users, including information about the value of water in all alternative uses, on the demand and supply conditions for every user, and if they could act immediately, then regulatory policies could be determined to ensure efficient resource allocation. Given that public authorities cannot by any means acquire such information at a reasonable cost, if at all, non-tradable water rights systems cannot achieve economic efficiency and equity. They are likely to result in an allocation which is rigid over time and unresponsive to changing social values.

> There is no way that a well-intentioned bureaucrat can know what constitutes a beneficial use without market transactions. It is the trading of well-defined and enforced property rights which will enable individuals with 'the knowledge of the particular circumstances of time and place' ... to coordinate their knowledge (Anderson, 1985).

Continuous trading in water rights generates prices that indicate the opportunity cost of water by co-ordinating dispersed information and preferences. A price is an information-rich signal which summarises all information available to market participants and motivates appropriate levels of individual action in response to changing demand and supply conditions. Thus, price performs the crucial rationing function in allocating resources to different uses and users. Transferable water rights, therefore, create a system of economic incentives in which those who have the best knowledge about the returns to water in its intended use – the water users themselves – are encouraged to use that knowledge to allocate water to higher-value uses. A system of transferable rights, thereby, maximises the economic value obtained from a scarce resource with a minimum of bureaucratic interference and apparatus.

Predictability

Markets are predictable in the sense that resources are reallocated through transactions that occur in response to changes in supply and demand. The flexibility sought through the market, however, reduces future predictability. In the market the future prices in water transfers, and hence the equilibrium distribution of water rights, are unknown. As a result, it is difficult, if not impossible, to anticipate how extensive the reallocation from one use to another might be.

While it is possible to introduce water markets anywhere without undue delay and this does not generally entail any major problems, it can be difficult and expensive, if not impossible, to reverse the process. Abolishing a market would mean that a government would have either to buy back the rights or expropriate them. The latter is likely to be politically unfeasible and, if accomplished, could possibly undermine investors' confidence in the economy, while the former would be complex and, almost certainly, a prohibitively expensive undertaking,

It is preferable that any innovation in water resources management be actively adaptive. New approaches, as radical as the introduction of a market, should seek to learn from experiences. Since surprise outcomes are to be expected, there is an argument that perhaps initial market trading should be conducted at a small scale and under supervision to minimise the chance of irreversible, adverse outcomes. A slow evolutionary process can be an advantage rather than a disadvantage. The initial steps should obviously be consistent with the final design of the system.

> Starting small gives both the institutions and the parties a chance to adjust and to become familiar with the system. Since most initial efforts will be precedent setting, it will take time to work them out. Once the precedents have been established, however, the process will become smoother, quicker and better able to handle a larger number of participants ... and trades (Tietenberg, 1995).

Experience in Chile suggests, however, that, by their nature, water markets may evolve very slowly and any concern of potential disruption in water management may not, therefore, translate into reality (Muchnick, Luraschi and Maldini, 1998).

The Fairness of the Market

Market transactions are fair in the sense that water reallocation takes place through voluntary mutually beneficial trades with perceived advantages for all the parties involved. Markets, however, can only guarantee fairness when no single participant can affect prices. Moreover, unless conducted in an institutional framework that causes participants to take into account third party impacts, markets generally cannot guarantee fairness to third parties that may be negatively affected by market transactions.

Since the future prices in water transfers and the equilibrium allocation cannot be known when a decision is made to introduce water rights transferability, the implications for income distribution cannot be known beforehand. There is no particular reason to expect that a water market will change income distribution in any particular way. If equity and other important collective, public or social values related to water use are an important part of water policies, such concerns can usually be accommodated within the logic of the market system. This can be done, for example, by purchasing water rights or reserving them in the initial allocation of rights (Cummings and Nercissaintz, 1992).

In general, concerns about equity should probably be treated outside the market. This is a problem of income distribution, not of the mechanism for water allocation and reallocation. Water marketing is unlikely to create 'new problems of unequal or unfair distribution beyond the reach of government policy' (Scott and Coustalin, 1995).

Theoretically, some goals for a water allocation system, such as predictability, equity and fairness, and the need to reflect collective, public or social values, could be better served by non-market institutions. The existence of these problems, however, 'does not necessarily call for a non-market alternative' (Anderson, 1982). Market 'failure' in some abstract sense does not mean that a non-market alternative will not also fail in the same or in some other abstract sense. The relevant comparison is between imperfect market solutions and imperfect administrative or political solutions, rather than between imperfect market solutions and the mirage of idealized administrative solutions. Administrative approaches to water allocation are characterized by their implicit reliance on 'the ability of the few decision makers within a centralized structure to act objectively, omnisciently, and responsibly in pursuit of the public interest' [which is never realized in practice] (Anderson and Leal, 1988).

Examples of the Value of Water Markets

The benefits of marketable water rights are not just theoretical or an illusion, they are confirmed by many empirical studies. The most significant evidence is provided from studies of experience with water markets in Chile, Australia, Spain and the United States.

Water markets in Chile

The introduction of water markets in Chile coincided with a major increase in agricultural production and productivity. It is reasonable to conclude that the introduction of tradable, and particularly secure, property rights in water made a noticeable contribution to this overall growth in the value of Chile's agricultural production. This increase occurred within an agricultural sector largely dependent on irrigation, with no significant increase in either the supply of water or the area under irrigation. There has been considerable private investment in improving the existing irrigation infrastructure both on and off the farm. The influence of water markets, however, cannot be fully separated from the effects of other economic factors, especially stable economic policies, trade liberalization and secure land rights. This notwithstanding, trading in water rights has succeeded in reducing the need for new hydraulic infrastructure, improved overall irrigation efficiency and has reduced the number of conflicts over water allocation. It also appears to have facilitated the shift from low-value, water-intensive crops to higher-value, less water-intensive crops (Rosegrant and Binswanger, 1994).

In addition, market transfers of water rights have produced substantial economic gains from trade in some river basins (see Table 3.2). These gains occur both in trades between farmers and in trades between farmers and other sectors. In the example shown, economic gains from trade are relatively modest in inter-sector trade, because water is being transferred from profitable farmers to urban drinking water supply, so that even though the financial gain to the farmer who sells is large, the overall economic gains of the reallocation are relatively small. This is because if, prior to the sale, water was being not used by its owner, and was not being taken from the river or canal, other farmers would have used it downstream.

Table 3.2 Economic analysis of gains from trade: Elqui and Limari Valleys

	Number of shares traded	Gains from trade (US $ per share)	
		Gross	Net of transaction costs
Trades with the water utility	298	675	658
Other inter-sector trades	63	1,160	1,139
Agricultural trades	351	934	839
Total/average	712	846	790

Source: Hearne and Easter (1995).

Water trading in Spain

In the Huerta of Alicante in Spain, owners of water rights must claim *albalaes* or tickets to be able to participate in each water distribution. The tickets, available in 12 denominations for a constant supply of water from 1 hour to one-third of a minute, are freely tradable through public auction and an informal market. A simulation model comparison of this system with those found elsewhere in Spain, where trading was not permitted, found that the market approach adopted in Alicante was the most efficient in terms of net increases in regional income (Maass and Anderson, 1978). The differences are not great when there are only moderate water shortages, but are significant in conditions of severe water shortage.

Another comparison of the effects on net increases in regional income for alternative short-run operating procedures for distributing irrigation water in the irrigation environments of Spain and the United States showed similar advantages for a market system. This study compared a composite of conditions in Murcia and Valencia, in Spain with those in the United States, using conditions in Colorado and Utah. The results indicated that of the procedures that do not depend on full seasonal storage, either markets or farm priorities are the most efficient. The farm priority procedure, however, is very inequitable and has been adopted, as a short-term response, only in severe droughts. On the other hand, a market procedure ranks high in equity. The results of this study show that markets were the most efficient of all the stream flow procedures considered. The results also showed that the conventional wisdom that the procedures 'that rank high in efficiency will do poorly in distributing income equally among beneficiaries while those that do well in distributive equality will be inefficient ... does not apply to a wide variety of conditions in irrigation agriculture' (Maass and Anderson, 1978).

Experience with markets in the United States

There are active markets in Arizona, California, Colorado, Nevada, New Mexico, and Utah in the western United States. Markets are an important water allocation mechanism in many areas and most studies show markets to be relatively efficient in allocating water. Transfer patterns in all the cases studied clearly indicate a movement from lower- to higher-value uses. Studies of these markets show that market decisions and prices generally reflect the opportunity costs of water in alternative uses and take account of those third party effects involving consumptive water users, but not instream flow and water quality values. It has also been found, however, that due to regulation 'water markets typically deviate substantially from the competitive market model, and prices may serve as only a rough approximation of the social value of additional water supplies' (Saliba *et al.*, 1987).

One example is the century-old rental market for irrigation water in the Northern Colorado Water Conservancy District and five major irrigation companies in the South Platte basin, which has resulted in substantial returns and allowed the avoidance of considerable losses in crop production. The rules and customs developed for using a market mechanism to allocate water through rentals and transfers 'make possible a better adjustment of the land-water relationship than is normally found in western irrigated agriculture. They might well serve as examples for other areas in adjusting for the varying needs of water users' (Anderson, 1961).

Similarly, a review of two decades of market activity in the Lower Rio Grande Valley of Texas found active water marketing practices with significant volumes of agricultural water having been transferred to municipal and industrial use (Chang and Griffin, 1992). Analysis of representative market transactions indicated that municipal benefits from water marketing far exceeded agricultural opportunity costs. The study estimates municipal benefits from trades at about US$ 5,000 to US$ 17,000 per 1,000 cubic metres compared with the lost water values to irrigators which range from US$ 249 to US$ 1,894 per 1,000 cubic metres under optimistic agricultural circumstances.

A government controlled market, known as the 1991 California Drought Water Bank, generated, in both physical and financial terms, perhaps the largest annual set of regional water trades to occur until that time in the United States and possibly in the world (Howitt, 1994). The bank helped alleviate extreme drought conditions across California in 1991 and subsequent years.

In setting up the bank, buyer participation was open to corporations, mutual water companies, and public agencies, except the California Department of Water Resources. Buyers were required to meet rigorous criteria to qualify as having critical needs. Sellers were assured that transfers would not affect the standing of their water rights and they would not form a basis for any loss or forfeiture of these rights. They were also assured that transfers would constitute a beneficial use of water and would not constitute evidence of waste or unreasonable use. To motivate early sales, purchase contracts contained a price escalator clause. This clause provided that if, by a specified date, the average price in similar transactions exceeded the prices in the contract by 10 per cent, the seller would be entitled to the higher of the two prices.

The 1991 bank bought over 1.0 billion cubic metres of water through 348 contracts for approximately US$ 100 million. The bank charged about US 14 cents a cubic metre for water delivered at the State Water Project's Sacramento-San Joaquin Delta pumping plant. The bank sold about 480 million cubic metres of water to 12 purchasers; some 80 per cent were for municipal and industrial uses. The Metropolitan Water District of Southern California, serving the city of Los Angeles, bought over half the water.

On the whole, particularly given the crisis nature of the programme, the bank was a success. It provided an effective regulated market that reallocated water to users with critical needs at minimum cost. It succeeded in moving California from a condition of drought emergency to one in which all critical needs were met. Negative economic effects were minimal, and, overall, the bank generated substantial gains for California's agriculture and economy. Bank operations also provided some benefits to fish and wildlife. The net financial benefits of its operations were estimated at US$ 106 million and it also had a positive net effect on employment resulting in a statewide gain of 3,741 jobs (Lee and Jouravlev, 1998).

PERMANENT VERSUS TIME-LIMITED WATER RIGHTS

Water rights can be permanent or time-limited. Theoretically, as long as the rights are freely transferable, either option is acceptable. The duration of the right, however, determines how easy it is to organise a market, at what level of transaction costs it will operate and, perhaps more importantly, the nature of incentives water users will face to invest in the development and conservation of water resources.

Permanent water rights are often preferred for two reasons. On the one hand, rights are homogeneous because they are all of the same duration, which simplifies market creation and reduces transaction costs. On the other hand, since protected rights to the use of water are a crucial element in promoting investment, a system of time-limited water rights is not likely to provide the necessary degree of security to promote long-term investment and planning. There are also difficulties related to potential legal and economic uncertainty in water management. A system of time-limited rights can be difficult to implement because of the rigidity of water distribution infrastructure, to say nothing of the political difficulties of terminating rights.

Every water right represents both an established business that relies upon water and many people, including employees, customers, backward- and forward-linked industries and so on, that depend upon the business. For such a business 'to take away its water, and thus its chance to exist, would be wasteful and terribly unfair... . Quite a lot of people will be outraged if the agency pulls the rug out from under the old firm' (Williams, 1985).

Property rights in water tend in most markets to be, like those in land, of perpetual duration, as for example in Chile and the western United States. In the western United States though, given the rules of abandonment and forfeiture for non-use, they can perhaps be more precisely described as 'of indefinite duration' (Trelease, 1974).

A system of time-limited water rights, however, can have its attractions. First, to the extent that water rights may be initially granted on the basis of historic use, the negative effects associated with this method of initial allocation would be mitigated under a system of time-limited rights. In particular, time-limited rights could reduce both the risks of large water users gaining market power and the need for regulation to ensure that such market imperfections are avoided. The limited duration of rights implies that alternative allocation methods can be established when the present rights expire (David *et al.*, 1980).

Time-limited rights can also facilitate future water management policy changes, for example the introduction of minimum or ecological flow requirements and the like, could be accommodated more easily. Permanent, or rights of very long duration, promote investments and allow long-term planning by holders, and thus encourage economically efficient decisions. They are, however, expensive, if not impossible, to recapture by the government whereas short-term rights may simply not be reissued when they expire (Eheart and Lyon, 1983). Once permanent rights are granted it is very difficult to reverse the situation created, particularly when the capacity to impose *ex post* conditions is limited.

Any potential administrative flexibility offered by time-limited rights, however, comes at the expense of a corresponding increase in uncertainty for the holder of water rights and can create tension between public authorities and water rights holders. Water markets depend on secure ownership rights. If a system of time-limited water rights is used, rights would have to be of sufficient duration to provide reasonable security to holders, to allow sufficient time to amortise capital investment and, also, to provide adequate incentives to invest. The problem then becomes: can longer terms be prevented from becoming permanent?

In the same way, attempts to impose *ex post* conditions on permanent rights to accommodate changing economic and social demands and conditions, if not implemented carefully, have the potential to generate uncertainty and to undermine the market. With rights of sufficient duration, a system of fixed-term rights could strike a balance between the need to provide security to investors and the need to provide flexibility by making the rights subject to periodic review. Very long terms, however, again would probably turn fixed-term rights into permanent rights.

One option, to address the trade-off between uncertainty and flexibility, might be to issue water rights in a staggered pattern or with a different duration (Eheart and Lyon, 1983). Both methods possess the disadvantage of creating heterogeneous rights which could make it difficult to organise a market and this would increase transaction costs.

If a right is to have time limitations, this must be clearly defined as a part of the right (Simpson, 1992). In addition, if the market is to allocate water rights efficiently, the duration of rights should not be subject to any particular actions on the part of the user (Eheart and Lyon, 1983). It must be recognised that limitations on the term of water rights, on the possibility of their renewal as well as other restrictions on the type of use that can be made of them lessen their value. Such limitations will also discourage transfers and reduce drastically the value of market allocation.

Time-limited water rights are used in Mexico where, under the 1992 water law, the right to use water is granted through 'concessions' to private individuals or corporations, or 'assignments' to federal, state and municipal entities. Both concessions and assignments can be granted for renewable periods of from 5 to 50 years. The average term has been more than 30 years to ensure security of the water rights (Rosegrant and Gazmuri, 1994). It can be suspected, however, that these inevitably will become permanent rights because of the inherent contradictions discussed above.

Rules of Allocation

The natural variability of any water supply affects the hydrological security of water rights, which in turn affects their market value. Rights drawn on more secure resources, such as perennial streams, groundwater aquifers or large lakes and reservoirs are more valuable. Any changes in the amount of water cause the yield of a right to vary. For this reason, there is a need to have allocation rules that relate available water supplies to individual user's permitted withdrawals or consumption.

Allocation rules are the institutional response to the need to make water rights relatively secure. The way in which allocation rules are defined determines how easy it is to organise a market for transferable water rights and how the risks associated with water shortages are shared among rights holders (Colby, 1988). While there are many ways to allocate water supplies, the major alternatives used in practice are priority allocation and proportional rules. In practice, water rights' systems often combine elements of both methods.

With a rule of priority allocation, water rights are defined in terms of two parameters, quantity and priority. Priorities may be determined in terms of time of use, type of use, or location. Defining priorities in terms of the type or location of use considerably hampers the flexibility that a market can provide in water allocation and in the transferability of rights. Consequently, the most common priority rule is that reflected in the prior appropriation doctrine of the western United States which operates on the 'first in time, first in right' principle. This means that, in times of water shortage, senior rights holders are satisfied first. Since the date at which a right was first obtained determines the relative reliability of the right providing water, where there is a market, earlier rights usually command higher prices (Colby, Crandall and Bush, 1993).

Where a proportional rule applies, water rights are defined in terms of a fraction of the available flow in the stream or canal, of the water available in a reservoir or lake or in terms of shifts or hours of availability at a certain intake. All water rights holders have equal priority sharing available water according to the proportion of rights held. Thus, all water users drawing water from the same sources share the insecurity inherent in variable water supplies. In Chile, although in the water law rights are defined in volumetric terms, in practice, proportionate rules are applied. Chilean rivers are divided into sections, and each point of withdrawal, at the moment of withdrawal, receives a percentage of the water in the respective section based on the number of rights held. Much of the work of the water users-organisations involves measuring flows and allocating the water corresponding to each right.

There is, however, some prior allocation influence in the allocation of water in Chile. A distinction is made between permanent and contingent or eventual water rights. Permanent water rights are allocated up to the average flow of a river and they have the first claim on available water. Contingent rights are granted for flows above the average flow and can be honoured only after all permanent rights have been satisfied. The allocation of water among contingent rights holders is governed by a priority system according to the seniority of the rights.

In some localities in Australia, holders of water rights where there is storage use the concept of capacity sharing to reduce uncertainty (Livingston, 1993). Water rights are defined as shares of reservoir capacity, allowing for inflows and reductions due to evaporation and seepage (Frederick, 1993). Water transfers are accomplished by making the appropriate debits and credits. Where adequate storage is available, capacity-sharing simplifies water transfers and allows water users better control over the timing of water deliveries. 'It is as if each user of water, or group of users, has their own small reservoir on their own small stream to manage independently from others' (Dudley, 1992).

The manner in which allocation rules are defined may affect the performance of the market, but deciding which approach encourages the most efficient allocation 'depends upon the specific characteristics of the water supply and users' (Eheart and Lyon, 1983). Theoretically, if a no-cost short-term water market were possible, there would be little or no difference between the systems. Trades under either system would tend to allow equivalent *ex post* outcomes to be attained (Eheart and Lyon, 1983; Howe, Schurmeier and Shaw, 1986b). In reality, however, markets do involve transaction costs and there can be other impediments to trading as well.

A priority rights system appears to have efficiency advantages in areas of mixed water use where water users are heterogeneous with regard to their demand functions and risk avoidance. In contrast, a proportional system appears to be advantageous in uniform systems where users' demand functions and risk avoidance are similar (Howe, Schurmeier and Shaw, 1986a, 1986b). Under a priority system, risk-sensitive water users can acquire more senior rights at higher prices, while those who are less sensitive to water shortages can acquire more junior rights at lower prices. Thus, market transactions allow the transfer of senior water rights to those economic activities which most value reliable water supplies. Water users can achieve the desired level of security and still only hold rights equal to the average amount of water needed. Any inefficiency during periods of shortage can be solved by short-term exchanges of rights (Howe, Alexander and Moses, 1982).

Under a proportional system, water users who need a reliable supply can reduce the probability of shortage only by holding water rights in excess of average needs. This practice can introduce inefficiencies and encourage hoarding. During periods of normal supply any rights held in excess of normal demand will usually be leased to other users. Such practices are relatively common in Chile where many farmers hold on to what may seem 'surplus' water to assure themselves of secure supplies in dry years (Bauer, 1995, 1997; Peña, 1996b). At other times, this 'surplus' water is either unused, benefiting other water users downstream, or leased, but it will not be available in dry years.

Another advantage claimed for a priority system is that, unlike a proportional system, it provides greater certainty of tenure (Burness and Quirk, 1979). For example, under a proportional system, the appearance of a new claimant to water reduces the certainty of tenure among existing water users, but seniority protects the privileges of existing users under a priority system. It can be expected, therefore, that the long-term average yield of a water right under a non-priority system will be less than the volumetric limit of the right. Under a priority system, however, the rights of senior holders will always be completely satisfied.

A priority system of water allocation has, however, the major disadvantage for the organisation of a market in the heterogeneous nature of rights. This makes it difficult for a market to function well, as it increases the costs of transactions. A proportional system, in contrast, has homogeneous rights. This undeniably facilitates market creation. This conclusion is supported by experience with proportional rights both in Chile, and in the Northern Colorado Water Conservancy District, which is perhaps the best example of a functioning water market in the western United States.

This difference between priority and proportional systems makes clear an important point about using a market system to allocate water. Markets work best where there is little overt interference. Markets may require regulation and rules to function efficiently and equably, but these rules and regulations should not require scrutiny of each individual transaction or create uncertainty about the nature of the good being traded. A priority system of water rights errs in exactly this respect and the advantages cited for it over non-priority systems disappear with greater hydrological knowledge.

REGULATING WATER MARKETS

The idea of treating water as an economic good, as private property, is so novel that using markets, rather than bureaucratic decision, for water allocation makes almost everyone responsible for water policy very nervous. Market allocation of water does not provide a perfect solution, even for fervent supporters, just a much better one. Any market requires rules and regulations. This should not be a surprise. Nor should it be a surprise that in moving towards market solutions all the issues cannot be solved in advance. The run of events will create unexpected situations; conflicts will arise that the existing rules maybe did not foresee. However, the majority of the difficulties that have been observed in those countries where water allocation is left to the market, can also be seen in countries with bureaucratic and political systems of allocation. Obviously, those countries reap none of the beneficial effects that the market provides for the vast majority of water allocation decisions.

The reallocation of water can have both positive and negative effects on those who are not party to the decision. The existence of such effects or externalities raises the possibility that any transfer can be beneficial to those immediately involved, but inefficient from an overall social perspective. To the extent that water transfers are associated with significant externalities, in a market, prices will deviate from the true opportunity cost of water, and hence will neither convey accurate market signals, nor encourage efficient water allocation decisions. Economic efficiency requires that all costs and benefits associated with use and transfer decisions be accounted for.

Externalities will emerge under any allocation system, but when water use and transfer decisions are decentralised to the level of individual users, as they are with water markets, the risks imposed may change. The decentralisation of water use and transfer decisions with a market system implies increasing the number of decisions. This fragmentation can introduce greater complexity in the control of externalities depending on the overall nature of the water management environment. The greater the user involvement in water management in general, however, the less likely it is that decisions to reallocate water can have unexpected repercussions on other users.

Externalities and Water Marketing

From the viewpoint of economic efficiency, if the market is to be used to allocate water or not, in any reallocation the holders of water rights must face the full opportunity costs of their actions. This requires that any external effects be accounted for. If institutional arrangements do not cause buyers and sellers to account for the external effects of their decisions, then the proper role for governments is to intervene to correct any external effects and to restore or substitute the requisite conditions for economic efficiency. Several policy options are open for correcting externalities, however they arise.

The problem could be left to voluntary resolution. Such a solution implies that, if there are external costs, society will sustain a welfare loss. There may be cases where this is an appropriate solution, particularly where any external losses are small both absolutely and in relation to the costs of regulation. This can be expected to be the case with the most common local transfers of water rights.

Conversely, water transfers might be absolutely forbidden in the presence of any external effect. Forbidding water transfers in the presence of externalities is, obviously, grossly inefficient because water allocation would remain locked into below optimum use patterns and many beneficial transfers could not take place, so efficiency losses can be very substantial.

Attempts could be made to try to internalise external costs by levying taxes on those creating negative externalities and paying subsidies to those creating positive ones. This is increasingly the policy used for water pollution control.

It might be more effective, instead of intervening every time externalities occur, to try to solve any externalities arising from water transfers. This can be done by clearly specifying and enforcing property rights, so as to establish a more effective institutional structure for negotiations in the market. For example, when there are disputes, water user-organisations, regulatory authorities and the judicial system routinely review market transfers. Water user-organisations, or the relevant public authorities, are typically the first arbiter. If the solution at this instance is not satisfactory, the dispute will be brought to the courts or eventually require a political solution. As in all disputes about changes in ownership and use of property, much litigation can be avoided if rights are clearly spelled out in the laws governing property and property transfers.

Common Externalities in Transfers of Water Rights

In the discussion of externalities, it is useful to distinguish between two broad groups of externalities that tend to be associated with transfers of water rights. First, there are externalities caused by physical changes in the location of water diversions or 'return flow' effects, such as changes in downstream flows with surface water transfers, and changes in the water table with groundwater transfers. Second, there are those caused instream, such as changes in fish and wildlife habitat, or recreation opportunities stemming from variations in streamflow or in water quality.

The first type of externality, due to changes in the physical location of withdrawals or the return flow effect, occurs because normally only part of the water withdrawn from a stream is consumed. The water that is not consumed will return at some point to the stream, either directly, by surface return flow, or indirectly, through groundwater. Once returned it can become subject to downstream appropriation, although this means granting rights to more than the base flow of the stream. Whenever return flows are used, any subsequent change, which alters the established pattern of return flows, can damage some users and benefit others (see Figure 3.3). Return flow effects can be significant, but there is often a time-lag before they become noticeable. Consequently, it can often be difficult to determine whether they are the result of the stochastic nature of river flows or of an upstream transfer and if the latter, it can be difficult to identify the transfer in question.

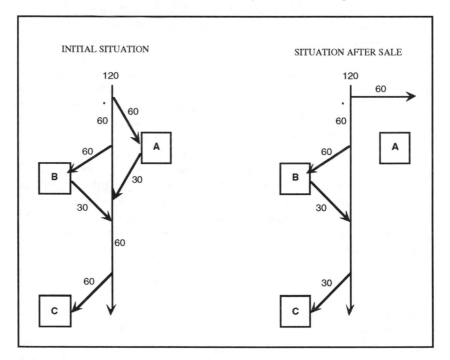

Source: Based on Holden and Thobani (1995).

Figure 3.3 Water transfers and return flow effects

The problem can be avoided by defining water rights as consumptive entitlements and not granting rights to return flows, as provided for in the Chilean 1981 Water Code. Alternatively, it can be done by defining entitlements at the source and partitioning run-off, storage capacity, and evaporation and seepage losses. The latter may be feasible, however, only in some circumstances, such as with the development of new water supplies. Under conditions of incomplete information about the water resources on which the rights are based, it can lead to difficulties where there are already formal and informal water rights based on return flows, as is the case in many countries, including Chile.

It can be assumed that in any long-established irrigated region there will be users making use of the return flows from upstream diversions. The introduction of water marketing may require greater attention to this problem because of the need to clarify rights, the development of a more intensive pattern of water reallocation, and because the transferability of water rights provides strong incentives for water users to use water more efficiently and to invest in water conservation, thereby reducing return flows.

Obviously, the existence of return flow effects are a possible source of inefficiency in any transferable water rights system, because a failure to take them into account may result in transfers in which social costs exceed social benefits. At the same time, attempts to guard against the losses to those relying on return flows may impede socially desirable transfers and introduce inflexibility in water reallocation by unduly increasing transaction costs. In the western United States, for example, protection against return flow effects is a primary cause of high transaction costs in most states (Williams, 1983; Saliba and Bush, 1987).

The possible options for minimising return flow effects from water reallocation include:

1. limiting trading to that portion of a water right actually consumed;
2. limiting trading to a given hydrologic area;
3. establishing property rights over return flows.

Whatever alternative is adopted, the regulation of return flow effects is complex and costly as it requires reliable and timely hydrologic data.

For example, the determination of historical use may be based on: actual records, if they exist, but they are rarely adequate; expert testimony, which is costly or; estimates of the amount of water that would have been required. In some places standard conversion rates are used to streamline the transfer approval process. For example, in New Mexico, the state administrative agency sets a standard quantity of water that may be transferred per unit of irrigated land. Anyone who disagrees with this quantity bears the costs of demonstrating that some other amount is appropriate (Colby, 1995).

Other jurisdictions in the United States have adopted an alternative policy. Water rights holders in a given geographic area or river basin surrender their individual water rights to the irrigation district or to a mutual water company, in exchange for shares entitling them to a specified quantity of water (Gould, 1989). Water rights can then be transferred freely within the entire service area by the transfer of shares and return flow effects are ignored. A variation of this approach is used by default in many parts of Chile where the restrictions in the water conveyance infrastructure limit transactions to users served by the same canal system.

In Chile, the narrow central valley, where most of the country's irrigated land is located, is divided from north to south into a series of fairly small, short and steep river basins separated by hills. This makes it expensive to transfer water between neighbouring basins or from downstream to upstream areas within the same river basin. Inadequate infrastructure for storage, diversion and conveyance also acts as a constraint on water transfers. As a result, transportation costs are often high except for transfers between neighbouring or nearby users on a shared canal system (Bauer, 1997).

Mainly because the Andean snowpack provides natural, short-term water storage, substituting for artificial reservoirs, Chile has few medium to large reservoirs for irrigation, and thus very little long-term storage capacity. Because of the lack of artificial storage capacity, most of the hydraulic infrastructure has been designed, built and operated for diverting water from unregulated rivers with highly variable flows.

The water flow in most canal systems is controlled by weirs, which divert water directly from the river to the head of the canal. In many systems, due to the variations in river flow, these are temporary structures rebuilt or repaired annually. The temporary nature of the structures allows some flexibility in water transfers but makes it difficult to quantify the diverted flow. In the canal systems that employ more permanent structures, problems arise because the works are built to divert specific proportions of the natural flow and are often hard to convert to different specifications (Ríos and Quiroz, 1995).

Within canal systems distribution works are also often inflexible, using flow dividers designed to distribute fixed proportions of changing flows. Water transfers, even from one farmer to another, require the modification of all intervening flow dividers to ensure that the water rights of other users are unaffected. Except for minor transfers, any changes are often prohibitively expensive. Any modification is even more complicated for outside trades.

Some systems do employ adjustable gates. These allow greater flexibility in water transfers, because water flows can be changed by simply lowering or raising the corresponding gates. Operating and monitoring costs are, however, higher than with less flexible infrastructure, principally because more flexible infrastructure requires more monitoring and control of user behaviour.

The observed levels of market activity are closely related to infrastructure availability with water markets being more dynamic and effective in areas with better infrastructure and well-organised water user-associations. In contrast, market transactions are less common in areas without storage and where large canal systems use fixed flow dividers (Hearne and Easter, 1995).

LIMITATIONS AND MARKET IMPERFECTIONS

One of the fears that is expressed in most discussions of water markets involves the degree to which markets may be imperfect. Obviously, in any allocation process water allocations may be affected by the unequal distribution of power among users. Any market will also be characterised by hoarding or speculation, which are legitimate market activities. In addition, most observed water markets are characterised by having few transactions, being considered therefore 'thin' markets. The traditional response to these fears has been to suggest that markets are not appropriate for allocating water or to only permit them with the adoption of severe administrative restrictions on their operation. There are strong reasons to believe that, particularly with intelligent regulation, markets can better cope with such problems than any alternative allocation system.

Market Power, Speculation, Hoarding and 'Thin' Markets

Market power could be exercised in a water market either directly through monopolistic behaviour, as by a price-setting seller or through monopsonistic behaviour, as by a price-setting buyer. Indirectly, but most unlikely, it could be exercised by the existence of the potential for some economic agents to use control of water rights to exercise market power in the output market for a product for which water is an input.

In any market, there will be the potential for some economic agents to influence market price levels. Where economic power is dispersed, however, no single market participant is likely to be in a position to exploit market power to undermine competition so as to gain unjustified advantage. Where individual buying and selling decisions have a major impact on the price, prices may no longer reflect marginal values and they may cease to provide the market signals necessary for efficient resource allocation. A large share of transferable property rights, however, does not necessarily mean having influence over the outcome in the market (Hahn, 1984).

In theory, but hardly in practice, large holders of water rights, generally only public utilities, might attempt to manipulate prices to improve their positions in the water market and to obtain 'excess' profits. For anyone to execute monopoly power, it would be necessary to actually withhold water, which is extremely difficult, if not impossible.

Conversely, if there are few buyers, a buyer with market power could follow a strategy resembling that of a monopsonist with the view to force the price down below the competitive level. In the water sector, some monopsonists could also be monopolists in the output market (for example, water-related public utilities). To counter the monopsonistic behaviour, sellers may decide to limit the number of actors on the supply side by participating in water markets at the level of one irrigation district as a whole (Gardner, 1990). In the United States, there are examples where farmers have negotiated jointly to ensure that all right holders have an equal opportunity to sell and that all sellers would receive the same price (Saliba, 1987).

Empirical studies of the impact of market power on the cost-effectiveness of transferable discharge permit markets, which have a similar, if not more concentrated, structure than water markets, show 'that market power does not seem to have a large effect ... Successful cartels are difficult to establish and maintain for any commodity...' (Tietenberg, 1995). In fact studies of emerging water markets in developing countries show them to be characterised by a great deal of competition (Rosegrant and Binswanger 1994).

Evidence from water markets in the western United States suggests that high-visibility buyers of irrigation water rights, such as public utilities and businesses which are dominant actors in the regional economy, tend to pay more for rights than other buyers (Colby, Crandall and Bush, 1993). The price premium may reflect various factors, but perhaps most important is the fact that such high-profile buyers are concerned with protecting their reputation. This means that they may be inclined to pay somewhat higher prices for water rights to avoid the costs of negative public perceptions and to mitigate the controversy that often accompanies large agriculture-to-urban water transfers.

The potential for the accumulation of market power is perhaps greater in basins where abundant water is available, accompanied by rapid economic growth. This will especially be the case if the original allocation of rights is free of charge and without any requirement that the water be put to a beneficial use. Where the supply is fully appropriated and there are many holders of rights and active water markets, the danger that monopolies or monopsonies will develop is relatively small (Howe, 1997).

The possibility of monopolising the market for water rights cannot be entirely ruled out but, for the most part, there is little danger of any single user dominating any basin to the extent that market competition is restricted. If market power posed a problem, however, the appropriate response would be the application of general competition and antitrust or anti-monopoly policies. Restrictions on transfers, such as individual limits on ownership, may help prevent the emergence of market power, but they will also limit the flexibility of water users. Moreover, they are difficult to implement and could become a stifling influence on development. Other measures can be used to deal with the issue of market power. These measures include imposing regulatory controls over transfers which seek to prevent the wholesale acquisition of available supplies without a demonstration of present or reasonably foreseeable need, the doctrine of 'beneficial use', or the imposition of taxes for holding a water right without developing it within a reasonable time. The problem with the latter solution is establishing the form and level of the tax and not unfairly penalising those who have justifiable reasons for holding rights.

The second aspect of the market power problem – the potential for some economic agents to use water rights to exercise market power in the output market – is of little practical concern. There are few, if any industries, where water is a significant input, although there is little or no possibility for substitution in some important uses. Available water supply sources may differ in their characteristics, but all of them can substitute for one another to some extent. As a result, most potential users have a wide range of alternative supplies of water and alternative technologies. Users can substitute labour, management, or capital for water in many uses. In addition, many goods and services produced in the water sector are tradable, either nationally or internationally, and have a wide range of substitutes. When substitutes are available, competition arising from the threat of losing customers to substitute products and services can discipline the conduct of rights' holders. The main exception could be hydroelectricity generation, which suggests the need to develop an adequate regulatory framework and to avoid uncompensated and unconditional allocation of water rights. If any user were to follow a strategy designed to raise costs for potential competitors, it would be a question of fact as to whether they were engaging in illegal monopolistic practices. This is precisely the circumstance antitrust laws are designed to deal with.

Related concerns about water markets include the possibility of speculation in water rights. Speculators are, however, essential ingredients in any market. Their participation can help deepen and widen the market, thereby facilitating an essential function of markets – the establishment of a 'going' price. Speculators, in any market, are constantly trading-off the present value of a future sale against the present value of a current sale. To succeed, therefore, speculators must show superior foresight in choosing their timing. While speculators can play a useful role in a market, it would be as inappropriate to leave important water management decisions solely to their forecasting abilities, as it is to the bureaucrats the market replaces.

The possibility that speculation might distort prices through unequal bargaining power or monopoly control cannot be ruled out in water markets. This is a particular danger where water rights are granted free of charge where there is abundant water and in basins which are undergoing rapid demand growth. In developed market systems with fully appropriated water rights by many water users, these fears are probably exaggerated

An effect related with speculation is hoarding. Hoarding will result in thinner markets and fewer trades than might otherwise have been possible, and hence may act to reduce the overall economic benefits of water markets. 'Hoarding is a response to risk and it intensifies the very problem to which it is a response' (Tietenberg, 1995). Reducing the risk associated with the availability and price of future water rights militates against the likelihood of hoarding.

Optimally, a market involves a relatively large set of transactions taking place continuously over time. Thin markets, markets in which trades are rare, are common, however, in water rights. In thin markets, prices must be negotiated case by case. In a thin market, search, information and negotiation costs may be very large. A thin market, where each transaction is unique and negotiated on a case-by-case basis, is likely to be less effective in setting a price that accurately signals the value of water and transmitting this and other information to market participants. In a market with few transactions, demand and supply conditions can change quickly resulting in price volatility which increases price risk, reduces incentives to engage in trading and encourages hoarding (Tietenberg, 1995). A small number of participants makes any market more susceptible to manipulation. It is easier for participants to establish and exercise market power leading to inefficiency and limiting the efficiency gains from water exchanges. Carefully designed bargaining rules can facilitate the efficient operation of a thin water market and reduce the efficiency losses due to strategic bargaining behaviour (Saleth, Braden and Eheart, 1991; Saleth and Braden, 1995).

MARKETS WORK

The arguments that have been presented here, and the empirical evidence presented, show that, where they have been allowed to operate freely, water markets work. Making them work well requires a clear understanding of the required institutional and legal framework. It also requires the establishment of clear rules and regulations governing exclusive property rights, the necessity for simple transfer mechanisms and the corollary of a minimum of bureaucratic interference in the market.

At the same time, it is not argued that the introduction of water markets is a simple universal solution to all the problems facing water management. It bears repeating that a water market is no more, and no less, than a management tool. It is a powerful tool, however, which spreads the burden and difficulties of management among a larger population, permits greater participation in management decisions and introduces greater flexibility into water allocation processes. At the same time, however, gaining the benefits of the more efficient water allocation that can be achieved with the establishment of a water market demands support from new skills and new attitudes in government. It will also require considerable investment by both the public and the private sectors in the registration of rights, in monitoring and measurement systems and in improving water distribution and transportation systems.

REFERENCES

Anderson, Raymond L. (1961), 'The irrigation water rental market: a case study', *Agricultural Economics Research*, **33** (2), 54–58.

Anderson, Terry L. (1982), 'The new resource economics: old ideas and new applications', *American Journal of Agricultural Economics*, **5** (4), 928–934.

Anderson, Terry L. (1985), 'The market alternative for Hawaiian water', *Natural Resources Journal*, **25** (4), 893–909.

Anderson, Terry L. and Donald R. Leal (1988), *Going with the Flow: Expanding the Water Markets*, Policy Analysis No. 104, Washington, DC: Cato Institute.

Bauer, Carl J. (1995), *Against the Current? Privatization, Markets, and the State in Water Rights: Chile, 1979–1993*, a dissertation submitted in partial satisfaction of the requirements for the degree of Doctor of

Philosophy in Jurisprudence and Social Policy in the Graduate Division of the University of California at Berkeley.

Bauer, Carl J. (1997), 'Bringing water markets down to earth: the political economy of water rights in Chile, 1976–95', *World Development*, **25** (5), 639–656.

Burness, H. Stuart and James P. Quirk (1979), 'Appropriative water rights and the efficient allocation of resources', *The American Economic Review*, **69** (1), 25–37.

Cestti, Rita and Karin Kemper (1995), 'Initial allocation of water rights in the United States, Australia, and Chile', *Infrastructure Notes*, March, No. WR-3, Transportation, Water and Urban Development Department, Washington: The World Bank.

Chan, Arthur H. (1995), 'Integrating equity, efficiency, and orderly development in groundwater allocation', in Ariel Dinar and Edna Tusak Loehman (eds), *Water Quantity/Quality Management and Conflict Resolution: Institutions, Processes, and Economic Analysis*, Westport, Conn.: Praeger Publishers, pp.143–153.

Chang, Chan and Ronald C. Griffin (1992), 'Water marketing as a reallocative institution in Texas', *Water Resources Research*, **28** (3), 879–890.

Colby, Bonnie G. (1988), 'Economic impacts of water law – State law and water market development in the Southwest', *Natural Resources Journal*, **28** (4), 721–749.

Colby, Bonnie G. (1995), 'Regulation, imperfect markets, and transaction costs: the elusive quest for efficiency in water allocation', Daniel W. Bromley (ed.), *Handbook of Environmental Economics*, Oxford: Basil Blackwell Ltd., pp.475–502.

Colby, Bonnie G., Kristine Crandall and David B. Bush (1993), 'Water right transactions: market values and price dispersion', *Water Resources Research*, **29** (6), 1565–1572.

Cummings, Ronald G. and Vahram Nercissaintz (1992), 'The use of water pricing as a means for enhancing water use efficiency in irrigation: case studies in Mexico and the United States', *Natural Resources Journal*, **32** (4), 731–755.

David, M., W. Eheart, E. Joeres and E. David (1980), 'Marketable permits for the control of phosphorus effluent into Lake Michigan', *Water Resources Research*, **16** (2), 263–274.

Donoso, Guillermo (1998), 'Análisis del funcionamiento del mercado de los derechos de aprovechamiento de agua e identificación de sus problemas', presented at the Seminario Institucionalidad y Gestión del Agua, Programa de Derecho Administrativo Económico, Universidad Católica de Chile, Santiago, Chile, 24–25 November.

Dudley, Norman J. (1992), 'Water allocation by markets, common property and capacity sharing: companions or competitors?', *Natural Resources Journal*, **32** (4), 757–778.

Eheart, J. Wayland and Randolph M. Lyon (1983), 'Alternative structures for water rights markets', *Water Resources Research*, **19** (4), 887–894.

Fox, Irving K. and O.C. Herfindahl (1964), 'Attainment of efficiency in satisfying demands for water resources', *The American Economic Review*, **54** (3), 198–206.

Frederick, Kenneth D. (1993), *Balancing Water Demands with Supplies. The Role of Management in a World of Increasing Scarcity*, Technical Paper No. 189, Washington, DC: The World Bank,.

Frederick, Kenneth D. and Allen V. Kneese (1988), *Western Water Allocation Institutions and Climate Change*, Discussion Paper Series No. RR88-02, Renewable Resources Division, Washington, D.C.: Resources for the Future.

Freeman, A. Myrick and Robert H. Haveman (1971), 'Water pollution control, river basin authorities and economic incentives', *Public Policy*, **19** (1), 53–74.

Gardner, Richard L. (1990), 'The impacts and efficiency of agriculture-to-urban water transfers: discussion', *American Journal of Agricultural Economics*, **72** (5), 1207–1221.

Gazmuri, Renato (1994), 'Chile's market-oriented water policy: institutional aspects and achievements', in Guy Le Moigne, K. William Easter, Walter J. Ochs and Sandra Giltner (eds), *Water Policy and Water Markets. Selected Papers and Proceedings from the World Bank's Ninth Annual Irrigation and Drainage Seminar, Annapolis, Maryland, December 8-10, 1992*, Technical Paper No. 249, Washington, D.C.: The World Bank, pp. 65–78.

Gould, George A. (1989), 'Transfer of water rights', *Natural Resources Journal*, **29** (2), 457–477.

Griffin, Ronald C. and Fred O. Boadu (1992), 'Water marketing in Texas: opportunities for reform', *Natural Resources Journal*, **32** (2), 265–288.

Hahn, Robert W. (1984), 'Market power and transferable property rights', *The Quarterly Journal of Economics*, **99** (4), 753–765.

Hearne, Robert R. and K. William Easter (1995), *Water Allocation and Water Markets: an Analysis of Gains-from-Trade in Chile*, Technical Paper No. 315, Washington, D.C.: The World Bank.

Hirshleifer, Jack, James C. De Haven and Jerome W. Milliman (1960), *Water Supply. Economics, Technology and Policy*, The RAND Corporation, Chicago: The University of Chicago Press,.

Holden, Paul and Mateen Thobani (1995), 'Tradable water rights: a property rights approach to improving water use and promoting investment', *Cuadernos de Economía*, **32** (97), 263–316.

Howe, Charles W. (1997), 'Protecting public values under tradable water permit systems: efficiency and equity considerations', paper presented at the *Seminar on Economic Instruments for Integrated Water Resources Management: Privatization, Water Markets and Tradable Water Rights. Proceeding*, Washington, D.C.: Inter-American Development Bank.

Howe, Charles W., Paul K. Alexander and Raphael J. Moses (1982), 'The performance of appropriative water rights systems in the western United States during drought', *Natural Resources Journal*, **22** (2), 379–389.

Howe, Charles W., Dennis R. Schurmeier and William Douglas Shaw, Jr. (1986a), 'Innovations in water management: lessons from the Colorado-Big Thompson Project and Northern Colorado Water Conservancy District', in Kenneth D. Frederick (ed.), *Scarce Water and Institutional Change*, Washington, D.C.: Resources for the Future, Inc., pp. 171–200.

Howe, Charles W., Dennis R. Schurmeier and William Douglas Shaw, Jr. (1986b), 'Innovative approaches to water allocation: the potential for water markets', *Water Resources Research*, **22** (4), 439–445.

Howitt, Richard E. (1994), 'Empirical analysis of water market institutions: the 1991 California water market', *Resource and Energy Economics*, **16** (4), 357–371.

Krutilla, John V. (1966), 'Is public intervention in water resources development conductive to economic efficiency', *Natural Resources Journal*, **6** (1), 60–75.

Lee, Terence and Andrei Jouravlev (1998), *Prices, Property and Markets in Water Allocation*, Serie Medio Ambiente y Desarrollo, No. 6, Santiago, Chile: Economic Commission for Latin America and the Caribbean.

Livingston, Marie Leigh (1993), *Designing Water Institutions. Market Failures and Institutional Response*, Policy Research Working Paper No. 1227, Washington, D.C.: The World Bank,

Lyon, Randolph M. (1982), 'Auctions and alternative procedures for allocating pollution rights', *Land Economics*, **58** (1), 16–32.

Maass, Arthur and Raymond L. Anderson (1978), *... and the Desert Shall Rejoice: Conflict, Growth, and Justice in Arid Environments*, Cambridge, Mass.: The MIT Press.

Muchnik, Eugenia, Marco Luraschi and Flavia Maldini (1998), *Comercialización de Los Derechos de Aguas en Chile*, Agricultural Development Network, Productive Development No. 47, Santiago, Chile: Economic Commission for Latin America and the Caribbean.

Pearce, David W. (1989), 'Economic incentives and renewable natural resource management', in *Renewable Natural Resources. Economic Incentives for Improved Management*, Paris: Organization for Economic Co-operation and Development.

Peña, Humberto (1996a), 'Conferencia del Ing. Humberto Peña', paper presented at the *Foro del Sector Saneamiento sobre el Proyecto de Ley General de Aguas*, 8–9 January, Lima.

Peña, Humberto (1996b), 'Water markets in Chile: what they are, how they have worked and what needs to be done to strengthen them?', paper presented at the Fourth Annual World Bank Conference on Environmentally Sustainable Development, Washington, 25–27 September.

Ríos, Mónica A. and Jorge A. Quiroz (1995), 'The market of water rights in Chile: major issues', *Cuadernos de Economía*, **32** (97), 317–345.

Rosegrant, Mark W. and Hans P. Binswanger (1994), 'Markets in tradable water rights: potential for efficiency gains in developing country water resource allocation', *World Development*, **22** (11), 1613–1625.

Rosegrant, Mark W. and Renato Gazmuri (1994), 'Establishing tradable water rights: implementation of the Mexican water law', in Mark W. Rosegrant and Renato Gazmuri (eds), *Tradable Water Rights: Experiences in Reforming Water Allocation Policy*, Irrigation Support Project for Asia and the Near East, Washington D.C.: The World Bank, pp.71–89.

Saleth, R. Maria and John B. Braden (1995), 'Minimizing potential distortions in a spot water market: a multilateral bargaining approach', in Ariel Dinar and Edna Tusak Loehman (eds), *Water Quantity/Quality Management and Conflict Resolution: Institutions, Processes, and Economic Analysis*, Westport, Conn.: Praeger Publishers, pp.385–397.

Saleth, R. Maria, John B. Braden and J. Wayland Eheart (1991), 'Bargaining rules for a thin spot water market', *Land Economics*, **67** (3), 326–341.

Saliba, Bonnie Colby (1987), 'Do water markets 'work'? Market transfers and trade-offs in the Southwestern States', *Water Resources Research*, **23** (7), 1113–1122.

Saliba, Bonnie Colby and David B. Bush (1987), *Water Markets in Theory and Practice: Market Transfers, Water Values, and Public Policy*.

Studies in Water Policy and Management, No. 12, Boulder, Colorado: Westview Press.

Saliba, Bonnie Colby, David B. Bush, William E. Martin and Thomas C. Brown (1987), 'Do water market prices appropriately measure water values?', *Natural Resources Journal*, **27** (3), 617–651.

Scott, Anthony and Georgina Coustalin (1995), 'The evolution of water rights', *Natural Resources Journal*, **35** (4), 821–979.

Seckler, David, David Molden, and Randolph Barker (1999), *Water Scarcity in the Twenty-first Century*, International Water Management Institute, Water Brief, No.1, Colombo, Sri Lanka: International Water Management Institute

Simon, Benjamin and David Anderson (1990), 'Water auctions as an allocation mechanism in Victoria, Australia', *Water Resources Bulletin*, **26** (3), 387–395.

Simpson, Larry D. (1992), 'Conditions for successful water marketing', in Guy Le Moigne, K. William Easter, Walter J. Ochs and Sandra Giltner (eds), *Water Policy and Water Markets. Selected Papers and Proceedings from the World Bank's Ninth Annual Irrigation and Drainage Seminar, Annapolis, Maryland, December 8–10, 1992*, Technical Paper No. 249, Washington, D.C.: The World Bank, pp. 97–201.

Simpson, Larry D. (1994), 'Are 'water markets' a viable option?', *Finance & Development*, **31** (2), 30–32.

Tietenberg, Thomas H. (1995), 'Transferable discharge permits and global warming', in Daniel W. Bromley (ed.), *Handbook of Environmental Economics*, Oxford: Basil Blackwell .

Trelease, Frank J. (1974), 'The model water code, the wise administrator and the goddam bureaucrat', *Natural Resources Journal*, **14** (2), 207–229.

Williams, Stephen F. (1983), 'The requirement of beneficial use as a cause of waste in water resource development', *Natural Resources Journal*, **2 3** (1), 7–23.

Williams, Stephen F. (1985), 'The Law of prior appropriation: possible lessons for Hawaii', *Natural Resources Journal*, **25** (4), 911–926.

Young, Mike (1997), 'Water rights: an ecological economics perspective', *Working Papers in Ecological Economics*, No. 9701, Centre for Resource and Environmental Studies, Ecological Economics Programme, The Australian National University.

4. Private Participation in Water Management

Privatisation or private participation in previously public sector activities is often conceived in terms of the sale of a publicly managed asset, a factory, a mine, an airline to a private investor. This is, however, a very narrow view of what private participation in any sector can be. Many assets held by the public sector cannot be sold for political reasons to private investors, some assets cannot be 'sold' for practical reasons, but their exploitation and administration can be undertaken by private agents. To take one extreme example, politically the government of Cuba remains opposed to private ownership of property, but it finds no difficulty with private companies managing its successful tourist industry; the assets, however, the hotels and land, remain state property. Practically, it is difficult to sell national parks, but in many cases these are now managed under private concession arrangements. We need, therefore, a wider idea of what private participation in water management can mean and not be restricted to the idea that the private sector needs to own the assets.

There are many forms of increasing the role of the private sector in the economy and of transferring activities or the rights to the net profit generated by an activity from the public to the private sector. Such transfers need not necessarily involve a change in ownership of the assets from the state to the individual. A change in the law may do as much or more to reduce the role of the state in the economy as the transfer of physical assets. For example, the removal of exchange restrictions or restrictions on the transfer of land received through a process of agrarian reform may completely change the state–individual relationship and greatly extend the area of the economy in which the market rules. In water management, as we have already discussed, the most significant act of privatisation may be the granting of property rights over water.

In many countries in the last few years, there has been a wholesale shift in the line dividing the public or state from the private or individual in nearly all aspects of society. The particular nature of the change depends on the previously prevailing situation, but, in nearly all countries, the shift has been significant. Water management, the water resource and the services based on water have not been excluded from this process. On the contrary, in many countries, the transfer of the responsibility for water-related goods and services and their management has formed an important part of the total process of widening the role of the individual and of the private sector in society.

The privatisation of the basic water services completely changes the demands on the water management institutions and also requires a thorough reconsideration of the policies that have been adopted towards water management in the past. Too often, the discussion of the role of privatisation in water management is limited by the inherited framework for water management. This can often considerably hamper innovations in structural arrangements which go beyond the mere transfer, in one form or other, of institutions from public to private management. There are many alternatives available for the incorporation of the private sector into water management and the experience of governments with these different alternatives will be discussed below. Considerable emphasis is given to the number and variety of alternatives available for structuring private participation. A lengthy discussion of these aspects is justified and of particular importance for the water sector. It must be remembered that the development of the idea or concept of water management has occurred within a context where the major users of water were within the public sector and where the private sector was largely excluded from participation.

The privatisation of water services forces a reconsideration and readjustment of the role of the state in water management. It demands not only that the state withdraw from many activities but also, that it take on new ones, often of a very different nature and requiring different skills and knowledge on the part of the public sector personnel. In water resources, all the experiences show that privatisation does not just stop with the transfer of assets, but requires continuing managerial actions within the public sector.

Water services, especially the provision of water supply and sanitation, tend to be natural monopolies. Natural monopolies pose a special challenge for public policy. There are, however, several options open to governments in confronting this issue, including:

1. A government might decide that monopoly rents are worth accepting and do nothing. Although this approach implies that society will sustain a loss in economic welfare, there may be cases where this loss is worth taking. Users may prefer paying monopoly prices for a high quality

service rather than going without or making do with an inferior supply. Moreover, where under-provision of services and their poor quality are the major problem, as in many developing countries, concerns about the imperfections of service provision by an unregulated private monopoly may be of little importance compared with the existing losses from poor provision. Furthermore, losses due to monopoly pricing by unregulated privatised natural monopolies are likely to be modest and at least some of them would be offset by the advantages given by size and integration.

2. A government might decide to continue the provision of the service through a public enterprise, although this option could lead to the continuance of the problems of efficiency, capital shortage, etc. which have prompted the reconsideration of such an alternative. There are, however, examples of reformed and efficient public enterprises.

3. Co-operatives are potentially an interesting option. There is considerable experience in many countries, both developed and developing, with autonomous, self-governing, voluntary co-operatives, particularly for small electricity or drinking water supply systems. However, co-operatives seem to work best only for smaller systems in rural areas and small towns. There is only one example in the world of a major urban water supply and sewerage co-operative, in Santa Cruz, Bolivia. Moreover, co-operatives, too, will require regulation.

4. A government might decide, as many governments have done in the last few years, to transfer services to private management and to use regulatory policy as a means of influencing private sector behaviour. This requires the establishment of an appropriate system of incentives to guide economic decisions in the private provision of water-based services under conditions of natural monopoly. In regulated utilities, the regulator acts as a substitute for the market, taking on some of the functions of competitors, attempting to provide similar incentives to improve efficiency by regulating aspects of the firm's conduct.

A further concern often expressed by governments who turn over water sector services or facility operation to the private sector is loss of control due to their inadequate regulatory capacity. When the public sector is no longer involved in the direct operation of a utility, it does not have the same control over its operations, for example, the level of production, quality, compliance with environmental standards, and so on. It also has no control over the contractor's inability to uphold the terms of the contract, such as unscheduled service interruptions or bankruptcy. If the partnership involves operation of a profitable utility, governments may also be concerned about the fiscal consequences of private sector participation.

THE GREATER EFFICIENCY OF PRIVATE MANAGERS

The experience of many countries, at all levels of development and over long periods of time, demonstrates that there is a clear and negative relationship between public ownership and political interference in an industry. Much of the blame for the poor performance of the public enterprises can be attributed to this political interference, with the consequent politicisation of key decisions and the lack of managerial autonomy, particularly decisions about prices and personnel administration.

One example of political considerations interfering with good management practice is the fact that many governments have heavily subsidised the provision of water-based services. This practice has led to the waste and overuse of water, misapplication of scarce resources, and serious distortions in the financial prices faced by producers and consumers which bias their production and consumption decisions away from welfare improvement. Usually tariffs bear little or no relationship to the cost of the services provided, the financial needs of the utility or to the consumer's willingness and capacity to pay. In addition, attempts to accommodate various special interest groups have led to complex and distorted pricing policies, taxation and regulations. These conditions, when coupled with frequent tariff and policy revisions, gave little incentive to use services efficiently or to reduce costs, but provide a strong incentive to exert political pressure.

It is commonplace to see state-owned companies being used to pursue goals that are unrelated to their entrepreneurial role. In many cases public companies have been used as vehicles for political patronage, corruption, nepotism, misappropriation of public funds, and indeed, as an instrument for furthering the political and material interests of the politicians in office.

Private ownership provides protection from undue political influences because it increases the transaction costs of government intervention in enterprise decision-making. The critical difference lies in the potential costs of intervention. Most important, in order to attract private capital, an institutional framework is required that protects private property and hence commits any government intervention to a mechanism that ensures non-negative expected returns. Secondarily, private ownership creates a separation between government regulators and the firm, making it relatively more expensive for the government to interfere.

Political interference and inefficient management are not inevitable in state companies; there are some examples of public water utilities working effectively. The problem is that politicians and governments find it difficult to commit themselves to good behaviour. The end result is that many countries have found it difficult to reform state-owned public utilities, except as part of a move to privatise them. Privatisation is increasingly seen as a way not only to increase efficiency but also to lock in the gains.

Under public ownership, property rights are attenuated as the public utilities and their managers and employees are not made the claimants of residual (profit) rights. They have, therefore, little incentive to operate efficiently. Moreover, public ownership is diffused among all members of society, and no one has the right to sell its share. There is, therefore, little economic incentive for any owner to promote the adoption of a more efficient provision of services. In contrast, private property rights are transferable and sellers of the rights can capture the capitalisation of efficiency gains. The ownership of private firms is concentrated among identified individuals, and thus the owners have incentives for continuously seeking improvements in efficiency and controlling management to ensure efficiency in the production of goods and services

The diffuse nature of public property also facilitates the capture of the benefits of public services by small and well-organised interest groups such as politicians, bureaucrats and other influential groups with an active role in the management and control of the services. This characteristic of public property gives these groups the capacity, opportunity and a strong incentive to usurp a disproportionate share of the service for themselves at the expense of the population at large.

In addition, public firms are expected to serve a variety of social and economic goals that are not always consistent. The mandate of the managers of state firms is often unclear and responds to multiple, often conflicting, demands. This gives rise to great ambiguity and creates incoherence in managerial decision-making, and also makes evaluation and monitoring of managers difficult. Private sector managers, in contrast, are more easily evaluated and better able to pursue efficiency because their objectives are more clearly articulated.

Private companies are subject to the discipline imposed by the private capital market and by the market for corporate control. Public companies, in contrast, tend to obtain their financing through the state. Even when they raise funds through the capital market, they have the explicit or implicit guarantee of the state. Thus the discipline imposed by the capital market is diluted.

Moreover, private sector managers often have a direct personal stake in the profitability of their enterprises, something that is normally lacking in the public sector where commercial objectives are subordinated to political goals and the threat of bankruptcy is absent. Public sector managers lack the financial rewards resulting from increases in portfolio values. Consequently, their planning horizon is short, commonly only until the next election, and the enterprises they manage are characterised by a shorter time-frame of reference, foregoing investments yielding longer-term returns in favour of short-term investments yielding immediate and visible benefits. The short planning horizon also makes them more vulnerable to shortsighted political pressures.

Finally, the largely administrative orientation of the public sector gives little reward for personal initiative and innovation. Rewards are based on longevity of service rather than on performance and contribution to the organisation's objectives: compliance with rules is given greater recognition than innovation. There may be specific disincentives for those who try to work hard, try out new ideas, and search for change and dynamism. In short, governments find it difficult to reward success adequately and to punish failure.

ALTERNATIVES FOR PRIVATE PARTICIPATION

A broad range of institutional alternatives exist for private sector participation in the water sector (see Tables 4.1 and 4.2). The options fall along a continuum between the extremes of almost completely public sector responsibility for management and service contracts through a degree of shared responsibility with concessions, leases, joint public–private arrangements to mainly private responsibility with divestiture. Obviously, these options, although different, may both overlap or be combined. Three broad alternative models, however, can be distinguished through which the private sector can participate in the provision of water-related services. These are full privatisation or divestiture, fixed-term franchises and other concession arrangements and everything else, including specific contracts, specially negotiated contributions and joint public–private arrangements.

Recent studies by the World Bank on trends in private participation in the water and sewerage and electricity sectors show that in the former divestiture is rare, but that it is usual in the electricity sector. The studies show that there is growing private participation in both kinds of utilities, but that it is still more common in electricity, especially generation, than in water supply and sewerage. The number of new private water supply and sewerage projects increased tenfold in the last seven years, but publicly financed projects still dominate and private investment is concentrated in only five countries (Silva, Tynan and Yilmaz, 1998). In contrast, private participation in the electricity sector is very widespread with more than 60 countries having

Table 4.1 Alternatives for private sector participation in water-related public utilities

	Divestiture	Concessions	Contracting-out	Joint arrangements
Independent capital funding	Full	Full	None	Part
Management control by the public sector	None	None	None	Part
Strategic control by the public sector	None	None	Full	Full/part
Risk to the public sector	None	None/some	None	Part
Need for regulation	Yes	Yes	Direct supervision	None
Efficiency incentives	Yes	Yes	Yes	Some
Improved competition	Limited	Limited	Yes	No

Source: adapted from OECD (1991).

made some progress in introducing private participation (Izaguirre, 1998).

Divestiture

The direct involvement of the public sector, other than municipalities, in the operation of water-related infrastructure is a relatively recent phenomenon. Until the late 1950s, the power sector in most countries was privately owned, and prior to 1950, it was common to see private provision of drinking water supply and sanitation services (Richard and Triche, 1994). The private sector has always maintained an important role in the development of irrigation in many countries. Today, however, it is regarded as radical innovation to sell the assets of public utilities or public infrastructure to private investors.

Table 4.2 Distribution of responsibilities under different forms of private sector participation

Responsibility	Ownership	Concessions and BOT	Leasing	Service contracts	Management contracts
Ownership of assets	Private	Public or mixed	Public or mixed	Public or mixed	Public or mixed
Investment planning and regulation	Private or mixed	Public negotiated with contractor	Public	Public	Public
Capital finance	Private	Private	Public	Public	Public
Working capital	Private	Private	Private	Public	Public
Execution of works	Private	Private	Public	Private as specified	Public or private as specified
Operation and maintenance	Private	Private	Private	Private as specified	Private
Management authority	Private	Private	Private	Public	Private
Commercial risk	Private	Private	Private	Public	Mainly public
Basis of compensation	Set by the regulatory regime	Based on terms of concession	Based on terms of contract	Based on terms of contract	Based on terms of contract
Duration in years	Indefinite	10 to 30	5 to10	Up to 5	3 to 5

Source: adapted from Kessides (1993).

The most commonly used methods of divestiture include the sale of shares, the sale of physical assets, opening a state-owned company to new private investment, and a management or employee buy-out. Sale of shares can be public or private. Under a public offering of shares, the state sells all or part of its holding in a company to the general public. Under the private sale of shares, a method more commonly used, the state sells to a pre-identified single purchaser or group of purchasers. A government may sell assets rather than shares. Again, assets may be sold individually or be sold together as a new corporate entity. Finally, a group of employees may acquire a controlling share in a former state company through a management or employee buy-out.

It is generally accepted, in the literature, that the transfer of public companies to private ownership can bring substantial improvements in productive efficiency. The findings of empirical research conducted by the World Bank and Boston University, in which 12 cases of privatisation were comprehensively analysed in Chile, Malaysia, Mexico, and the United Kingdom, indicate that privatisation does bring substantial gains (Galal *et al.*, 1994). Most of the enterprises in the sample were monopolies or oligopolies. Even so, in 11 of the 12 cases, the gains were positive and large, amounting to an average 2.5 per cent permanent increase in national income.

This empirical evidence is supported by several theoretical arguments which suggest that divestiture could be an attractive option for the achievement of greater efficiency. Although none of these arguments unequivocally implies that privatisation will significantly increase productive efficiency, some improvement is likely to result.

In discussing the advantages of privatisation, it is useful to distinguish two broad groups of industries. There are those industries which could, and in many cases do, operate in competitive product markets free from substantial market failures and those industries which cannot operate in competitive markets or exhibit market failures.

Overall, evidence suggests that privatisation of industries operating in competitive areas free from substantial market failure generally leads to significant efficiency gains and that private ownership is preferable on efficiency grounds. The studies suggest, however, that competition and market structure may be a more important influence than ownership and that the benefits of competition can overcome the tendency to inefficiency resulting from public ownership. Studies of contracting-out of publicly financed services previously performed by the public sector to the private sector, which implies an immediate increase in competition, reach similar conclusions (Stevens and Michalski, 1993).

One example is provided by the privatisation of GENER, an electricity generating company in Chile. The ownership of GENER was transferred to the private sector in 1987. The company operated with essentially the same generating equipment and under the same regulatory environment both before and after privatisation. A recent analysis of the welfare consequences of the privatisation concluded that the gains in productivity were large enough to make the divestiture of GENER welfare-improving by an amount equivalent to 21 per cent of the value of the company at the time of privatisation (Galal, 1992).

Privatisation of industries, which do not operate in competitive markets or have substantial market failures, is growing, in highways, for example, but experience with the impacts of ownership transfers is mixed. Some studies give the advantage to public ownership, others to private ownership, and yet others fail to find any significant difference between the two. Empirical studies of the relative performance of public and private enterprises have been undertaken in the United States, where the two types of ownership coexist in similar market conditions, but, again, there are conflicting opinions as to whether water utility efficiencies vary systematically with type of ownership.

It is in the tradable goods industries operating in competitive markets that the arguments for divestiture are strongest. In such industries market liberalisation and restructuring can be counted on to supply the beneficial pressures of competition. There will be only limited need for the more detailed and intrusive forms of regulation. In the water sector, this would include many irrigation projects and electricity generation. These activities produce outputs for which there is a wide range of substitutes. In tradable goods industries divestiture will have a positive impact on economic efficiency and may be welfare-improving. Divestiture enables benefits to be realised from the strengthening of private rationality under competitive pressure from both the product and capital markets. This is why so many governments have withdrawn from the electricity generating industry and transferred irrigation management to farmers.

Divestiture need not be complete to reap the benefits of private ownership. Joint public–private arrangements where the private sector owns a minority share of an enterprise can lead to highly positive results, especially if it is combined with other measures to improve efficiency and reduce political interference.

The experience of privatisation in many countries suggests that strategic investors do not insist on majority ownership of the companies whose shares they acquire. They do insist that adequate contractual arrangements are in place to give them management control and to prevent undue interference by the state as majority shareholder (Ahmad and Mainster, 1995). Under specific conditions, investors are prepared to acquire minority stakes even in companies where they do not have management control, provided that they are profitable and that there is no undue government interference in the day-to-day running of the companies.

In addition to being a good revenue-generating exercise, this method is particularly appropriate for dealing with funding problems of under-capitalised enterprises. It is also appropriate when the objective is to strengthen an enterprise which the government intends to keep in the state sector (Vuylsteke, 1988). The sale of a minority share of a company can provide positive incentives for efficiency and have a powerful behavioural effect on the firm's performance. It fosters independent decision-making and compels management to be more accountable. Furthermore, in order to be listed on the stock exchange, a company may have to introduce important changes in internal operations to comply with legal, financial and disclosure requirements. Such changes will be governed by the applicable laws of the country of offering and usually enforced by a securities and exchange commission or similar agency (Vuylsteke, 1988). Moreover, because the stock price is quoted daily, its managers would want to be leaner, cleaner, and meaner in order to demonstrate to the public that their company is a profitable investment (Nellis and Roger, 1994). Private minority ownership or minority control can be very useful as interim arrangements prior to more comprehensive forms of private sector participation. Even selling very small proportions of companies can be beneficial.

Joint public–private arrangements provide innovative means of co-operation between the public and private sectors with the potential to improve efficiency through the introduction of more business-style management and in the assessment of project risks and market feasibility. It can also lead to the injection of private sector capital, particularly when the private partner can readily raise bonds or issue notes, and has know-how and expertise, for example, in raising project finance.

Joint public–private arrangements also enable the public sector to act in a more flexible manner than if a service were entirely in the public domain. It also permits the public sector to play a role which it cannot usually adopt, for example, relaxing standards or earning profits to raise funds for investment in infrastructure or for direct service provision (OECD, 1987). In addition, joint public–private arrangements may provide a means of avoiding public sector borrowing controls. Finally, this approach may be attractive when it is not feasible, for political reasons, to transfer assets or full responsibility for service provisions to the private sector (World Bank, 1995).

A possible shortcoming is the risk for governments of conflict of interest problems, particularly if it is simultaneously regulator and owner (World Bank, 1995). In the case of partial privatisation, governments can be tempted to apply internal regulation instead of external regulation (Bös and Peters, 1988). When there is a lack of separation between regulatory and operational activities one of the most essential principles of regulation is violated. This usually results in a great deal of inefficiency because governments usually find it irresistible to meddle with day-to-day administration of the utilities.

On these grounds, leases, concessions or full divestitures, which provide for the separation of responsibilities, have considerable advantages.

The only existing completed major divestiture of drinking water supply and sanitation services is that in England and Wales. The British government decided on sale as the best means of transferring responsibility for drinking water supply and sewerage to the private sector. In its White Paper on the privatisation of the water industry in England and Wales, the British Government set out a number of arguments to justify the transfer. These arguments included:

1. freeing the companies from government intervention in day-to-day management and from political pressures;
2. releasing the companies from the constraints on financing that public ownership imposes;
3. providing access to private capital markets to encourage the pursuit of effective investment strategies;
4. giving the financial markets the ability to compare the performances of individual companies with other companies to provide the financial spur to improved performance;
5. designing a system of economic regulation to ensure that the benefits of greater efficiency are passed on to customers;
6. privatisation would also provide a clearer strategic framework for the protection of the environment;
7. private companies would have greater incentive to ascertain the needs and preferences of customers and to tailor their services and tariffs accordingly;
8. private companies would be better able to compete in the provision of various commercial services;
9. private companies would be better able to attract high-quality managers;
10. an opportunity would be provided for wide ownership of shares;
11. most employees would be more closely involved with their business and more motivated to ensure its success.

Similar arguments have also convinced the Chilean government to begin a major divestiture programme of its state-owned drinking water supply and sanitation companies.

In both countries, the reasons for privatisation were partly political and partly financial, particularly the huge investment requirements and the need to reduce the financial contribution of the water industry to the public sector borrowing requirement. Nevertheless, the forms of divestiture chosen both in England and Wales and in Chile have some features in common with franchising. The companies hold a concession, called an 'appointment' in

England and Wales, for their area of responsibility. In England and Wales each appointment runs for a minimum period of 25 years and may be terminated by the government at any time on or after the end of that period, provided at least 10 years' prior notice has been given. In Chile the concessions are indeterminate, but can also be terminated for failure to comply with the regulatory rules. On this view, the force of the regulatory work in the industry may depend in practice on how realistic any risk of appointment revocation appears to the operating companies (Kinnersley, 1990). In both countries, the concession arrangements allow ownership to be transferred directly with the approval of the regulator.

The benefits from divestiture, despite the example of the success in England and Wales, however, are generally more difficult to realise in industries which do not operate in competitive markets or have substantial market failures. Transfers of natural monopolies, such as water supply, to the private sector, insofar as the transfer process will not result in effective and undistorted competition, call for permanent and detailed public regulation, as has been established in England and Wales and in Chile. Regulation, as will be discussed in Chapter 5 in detail, is intrinsically imperfect, largely because of the basic informational asymmetry between the regulator and the firm. These considerations help explain, at least in part, why, in spite of the fact that divestiture of public enterprises has become more frequent practice, the World Bank database shows few examples of the transfer of drinking water supply and sewerage systems to private ownership.

Concessions and Related Franchising Arrangements

A concession is an agreement in which a public authority awards to a private company, usually through a competitive qualification process, a fixed-term right to provide a service with characteristics of a monopoly within a defined geographical area. There can also be non-exclusive, competitive concessions. It is by far the most common form that has been adopted in the water supply and sewerage sector (Table 4.3). In electricity, however, private ownership, whether through divestiture of existing plants or new 'greenfield' projects, is far more common, accounting, according to the World Bank, for over 90 per cent of all private participation in this decade (Izaguirre, 1998).

Concessions or franchising are attempts to harness market forces and to increase efficiency through an auction of the right to operate a natural monopoly. The main arguments for the use of franchising rather than direct sale is that concessions may reduce the need for the more intrusive forms of regulation and yet ensure that a natural monopoly does not charge a

*Table 4.3 Private participation in water and sewerage projects in developing
countries, by type, 1990–1997*

Type	Number of projects	Total investment (US$ millions, 1997)
Concessions	48	19,909
Greenfield	30	4.037
Divestiture	6	997
Operation and management contracts	13	N/A
All types	97	24,950

Source: Silva, Tynan and Yilmaz (1998).

monopoly price. Franchising encompasses leasing and other institutional arrangements for private sector participation as well as concessions.

Franchising can introduce the efficiency characteristics and mechanisms associated with free markets into natural monopoly situations where direct competition is not possible. In a natural monopoly situation only one firm will actually produce a good, even though many firms are capable of doing so. In order to exploit this possibility for competition among the firms that could produce in the industry, bids are opened for the operation and provision of service and the winner becomes the monopolistic supplier. Obtaining a monopoly under such conditions becomes, itself, a competitive activity (Posner, 1975). The argument is that competition for the market among the potential producers will hold in check the potential monopoly power of the winner through the competitively determined terms of the concession. This competition should, at least in theory, increase efficiency and bid down price of the product to the point where it does not reflect the monopoly power of the eventual concession holder.

Franchising can also be justified on the grounds that with concessions and similar contractual forms it is possible to isolate those areas within a public utility where market failures are significant and contract for them. Other areas within a public utility, where competition is possible, can be privatised. Arguments can also be advanced for using a mixture of franchising or concessions and total divestiture or sale to reduce the burden of regulating totally privatised public utilities with natural monopoly characteristics.

Firstly, market failures are not pervasive in all aspects of water services, but rather are associated with some specific elements of service structure. By isolating activities with natural monopoly characteristics, their damaging consequences can be quarantined and competition can be introduced in the market segments where it is possible and desirable. This means restructuring so as to separate functionally the potentially competitive activities from those which are natural monopolies. Once restructured, the now independent potentially competitive activities can be opened to appropriate forms of competitive provision.

In many countries, water utilities have been characterised by complete vertical integration to the degree that they include all operational and support activities. Many utilities could realise substantial cost savings and enhance efficiency through separating out potentially competitive activities and leaving their supply to the market.

Secondly, the literature on 'contestable markets' (see, for example, Baumol, Panzar and Willig, 1982) holds that in a contestable market competitive pressures supplied by the mere and perpetual threat of entry can enforce good conduct by incumbents. Consequently, this will preclude excessive profits and prices as well as waste and inefficiency (Baumol and Lee, 1991). A contestable market is a market where entry and exit are without cost. A market can remain highly contestable if an entrant can achieve contractual relations with prospective customers, which render it immune from incumbent's retaliation. Where this is true, the incumbent can foreclose entry opportunities and protect its market share only by behaving well all along, that is providing customers with all the benefits that an entrant could be expected to bring.

In the water sector, public authorities can use concessions and other franchising arrangements to amplify the areas of contestability, thus easing the regulatory problem. Contestability can be facilitated through encouraging new private operators to enter the market by ensuring that they can compete on fair terms with existing operators.

It is argued that fixed-term concessions can offer advantages over divestiture, by providing a means to institute regulation gradually. This can be a particularly important factor in countries with little experience in formal regulation. In practice, however, concessions can create as many difficulties for the regulator as divested companies. It is also argued that franchising widens the possibility for incremental private sector participation. One advantage of an incremental approach is that market forces can be introduced without having to set up a completely new regulatory framework. Uncertainty about the future regulatory framework and other risks can be addressed through a series of long-term contracts between the government and private companies (Liétard and Santos, 1994).

It is further often argued that franchising also reduces the need for the most intrusive forms of regulation. In comparison with divestiture, for which a strong centralised regulator is considered necessary, it is argued, franchising can offer scope for regulation to be both more localised and less complex. This is because, basically, the degree of private ownership is less complete, and the possibilities for loss of the franchise are greater than for loss of ownership (Kinnersley, 1990). In addition, competition between informed potential franchisees might reduce the problem of asymmetrical information when granting a franchise.

It is important to note, however, that franchising does not eliminate the need for regulation. The franchise mechanism has important limitations. Unfortunately, many water management activities are particularly prone to such difficulties. Therefore, concessions usually entail an important degree of continuing regulation. The use of concessions and similar contractual arrangements should be seen, not simply as an alternative to regulation, but can be used as a means of harnessing some of the information and incentive advantages of competition so as to reduce the burden on the regulator.

It is possible to conceive of concessions as interim arrangements in the transition to more comprehensive forms of private sector participation. During the term of the concession, the holder has an opportunity to assess the viability of the utility and could then make a purchase offer based on better knowledge of the utility's financial situation and potential (Bouin and Michalet, 1991). This is a very important consideration given that in most developing countries there is a general lack of knowledge about the conditions of the existing asset base and of patterns of consumption of public utility services. The downside is that the incumbent will have a considerable advantage over other potential bidders and also over the regulatory agency, particularly if it does not divulge all the available economic and financial information about the company.

Franchising has been particularly the practice in the provision of water supply and sanitation in France. In France, the municipalities traditionally have provided the water supply and by delegation, under a wide variety of contractual arrangements, through a private operator. The *Code des Communes* or municipalities stipulates that water distribution and wastewater disposal and treatment are municipal, industrial and commercial public services. The municipalities possess, however, a wide degree of flexibility in the selection of contractual arrangements. They can provide services directly, *régie direct*, or by delegation to a private operator under a variety of contractual arrangements. Today private companies provide drinking water supply services to about 75 per cent of the population and sewerage services to about 40 per cent (Chéret, 1994).

Under the French system, delegated management does not involve the transfer of assets; these remain the property of the municipality, even when the private operator finances them. The most common arrangements are management contracts, *gérance* or *régie intéresse*, leases, *affermage*, and concession contracts. The difference between a *gérance* contract and a *régie intéresse* contract is that under the former, the operator receives a lump sum payment and under the latter, the operator receives both a lump sum payment and a payment based on results. Leases and concessions, however, have gradually replaced both *gérance* and *régie intéresse* contracts.

The most widespread is the lease contract, under which the operator is responsible for management while the municipality is responsible for investment. Lease contracts are of relatively short duration, up to 12 years while a concession contract is usually for much longer. Under a concession the holder is responsible not only for management, but also for capital investment. In France, concessions are more common in drinking water supply and leases in sewerage. Under both arrangements, the initial contracts are generally awarded following a call for tender (Haarmeyer, 1994).

The system has created the most experienced private operators in the world. Three companies control more than two-thirds of the drinking water supply market in France: the Compagnie Generale des Eaux has 39 per cent, Lyonnaise des Eaux has 21 per cent, and Saur of the Bouygues group has 10 per cent. Although there is often fierce competition for the initial right to a franchise, the incumbent franchisee often wins contract renewals and it is very rare for an incumbent franchisee to be displaced. When the contract reaches the expiration date, its extension, with any modifications, is generally renegotiated with the current operator. The contract may be terminated by the municipality, but this is extremely rare (Kay, 1993). The fact that the municipalities retain the option of taking over the operation creates a margin of competitive pressure.

There is no formal regulatory system. The central government, however, reviews the contracts. Tariffs are determined on the basis of a forecast operating statement submitted by the operator in support of the bid. The contracts may include a formula for tariff adjustment on the basis of price indexes for salaries and social security charges, energy, chemicals and other items. There are also provisions for periodic reviews and procedures for certain agreed-on costs to be passed on to the municipality.

A key feature, which gives the franchise system in France and elsewhere some of its competitive characteristics, is that the assets of the sector will always revert to the public sector. Thus the options for choosing a different private operator or of direct public sector management will open at intervals (Kinnersley, 1990). There is also competition among the various management options under which the responsibility for service provision can be transferred to the private sector (Haarmeyer, 1994).

Characteristics of alternative franchising arrangements

Franchising arrangements are now widely used for the provision of the public services in many countries. They vary from the relatively simple contracting-out of specific tasks to the long-term concession of the total provision of the service, including responsibility for capital investment.

Contracting-out The contracting-out of an operation through service contracting or subcontracting is a transfer, by means of a fixed-term contract, of responsibility for specific services or elements of infrastructure operation and maintenance. Service contracts are usually fairly limited in scope and cover specific activities, such as meter reading or equipment maintenance. The contractor in these cases is paid for service delivery. Compensation may be based on a variety of methods, such as cost-plus, fixed-fee, lump sum, or unit costs, on a time basis or a percentage or proportional share of revenues. The fees are usually not directly linked to operational efficiency or cost control.

The public utility retains overall responsibility for the system, except for the specific services contracted out, and it finances working capital and fixed assets. The utility bears the full commercial risk for service provision. Control is exercised through setting performance indicators, detailed performance specifications and procedures for monitoring quality, evaluating bidders, supervising contractors, applying contract sanctions, paying an agreed fee for the services, among others.

Service contracts are usually negotiated for relatively short periods, normally less than five years, but can be even shorter. A longer duration is important for services that require substantial initial investment, for example, where specialised equipment must be bought. In such cases, the contract must allow for capitalisation and for the depreciation of capital expenditures, although such provisions should be carefully weighted against monopoly potential if contracts are for a long time period.

The potential benefits of service contracts include cost savings and efficiency improvement, better access to technology, equipment and expertise whose acquisition cannot be justified due to insufficient levels of use, the possibility of adapting operation and maintenance systems to varying demands, and so on. Service contracts are particularly appropriate for occasional demands, such as studies, engineering designs, construction, or when in-house demand is too small for efficient scales of production (World Bank, 1995).

Many activities can be contracted out. It is very common for auxiliary activities (such as cleaning, food catering, security and vehicle leasing), administrative, commercial, training, technical assistance and standard professional services (auditing, accounting, procurement, legal matters, payroll, data processing, such as preparation of optimisation models for reservoir operation, recruitment, and the like) and for managing non-core assets and activities. Contracting-out of non-core activities allows the company management to concentrate its efforts and resources on core issues of business and lets secondary concerns be taken care of by specialised companies (Nellis and Roger, 1994). Care needs to be taken in separating core activities, such as water production, treatment and distribution, and wastewater treatment, to ensure effective co-ordination, controls and supervision. Contracting-out of activities, which require close co-ordination and quality control, tends to be more demanding but is also possible and even advisable if adequate monitoring and co-ordination can be ensured. Services not appropriate for contracting-out, in themselves, can often be divided by function to permit their subcontracting

Contracting-out is commonly used by both public and privately owned water utilities as it is with organisations of all types. It is normal practice except in the most exaggerated and restricted of centralised economies. Most water utility operations can be and are contracted out, including construction, billing and collecting, meter reading, and operation and maintenance. Contracting-out can result in considerable savings. For example, in Bogota·, Colombia, it has been reported that it costs five times as much for the public agency to read a water meter than it costs when it is contracted out to private meter readers (Briscoe, 1993).

Traditionally, everywhere contracting-out has been normal practice for the design and construction of major capital works given the obvious benefits of specialised engineering knowledge and construction skills. Contracting-out of maintenance has also long been an established practice. In some cases, practically all of the core functions of public agencies can be contracted out to the private sector leaving the agency with only a basic staff to award and monitor the contracts (Kessides, 1993).

In India, Madras Metro Water has contracted services ranging from the provision of staff cars to the operation and maintenance of sewage pumping stations. EMOS, the water utility in Santiago, Chile, has contracted out services accounting for about half of its operating budget, including computer services, engineering consulting services, and repair, maintenance, and rehabilitation of the network. To enhance competition, the Santiago utility has at least two service contracts for each kind of task (World Bank, 1997). Contracts are usually awarded for one or two years under competitive bidding for meter reading, billing, system maintenance, vehicle leasing, and other activities (Easter and Hearne, 1993). As a result, EMOS has the highest staff productivity among Latin American drinking water supply and sewerage companies, even when an imputed labour cost for contracted services is taken into account (Yepes, 1990).

Management contracts   Under management contracts, while the owner retains full ownership and is responsible for capital expenditures, maintenance, and working capital, a private firm supplies management and technical skills. Sometimes, however, the manager may take an equity position. Unlike service contracts, which tend to be fairly limited in scope and tend to cover only one specific activity, management contracts transfer full managerial control, with the freedom to make day-to-day management decisions. Management contracts run usually from three to five years, although they can be longer and are often renewable. Management contracts are less frequent than concessions and only account for 13 per cent of private participation in projects in the 1990s, although in Africa seven out of ten projects are management contracts (Silva, Tynan and Yilmaz, 1998). Management contracts have become very common in recent years in the United States and more than 1,000 municipal systems are now managed under such arrangements.

A manager may be an individual or group of individuals with the skills and expertise required for the operation of the enterprise. Usually, however, it is a company from the water sector that offers the service. The scope of the management company's role can include activities such as general management, financial administration, personnel administration, production management and technology transfer, staff training, and marketing and distribution. Under a contract, the private manager may have wide powers over existing personnel, although they commonly remain employees of the original enterprise and subject to government pay scales and conditions. The success of a management contract is dependent on a number of factors including the viability of the project. It is also necessary that there is a supportive external policy environment. The owner must also be willing to support and delegate to the manager sufficient control and authority to manage the enterprise, including the key functions that affect productivity and service quality (Hegstad and Newport, 1987).

A large variety of compensation packages can be used, including annual fixed fees, a fixed fee plus costs, a fee as percentage of profits, sales or production, incentive payments based on increased production, profitability, input conservation, plant and equipment efficiency and availability, and the achievement of localisation targets.

The purpose of management contracts is often to acquire the contractor's expertise and knowledge while preparing a company's staff to run the operation. Basically, the management contractor's assignment is to perform certain functional responsibilities related to the operation of the project and to make certain that its corporate resources and skills are available to the enterprise during the contract period (United Nations Centre on Transnational Corporations, 1983). The contractor should, therefore, take full operational responsibility and assume the major part of the commercial risk.

A management contract provides a legal framework to achieve these aims while allowing retention of ownership and the overall setting of policy and budgetary control responsibilities by the owner. It can be a means of increasing efficiency in operations prior to moving towards fuller private participation. There are drawbacks including a loss of day-to-day operational control and the need to develop a regulatory system so as to monitor the performance of the manager to ensure that overall policy direction is maintained. Management contracts can be difficult, time-consuming and expensive to design and structure properly as well as to implement, both absolutely and relative to other options.

Management contracts do represent an attractive option where institutional capacity is weak, which explains their popularity in Africa. They also can be useful in cases where reforms of the regulatory framework for the sector are underway, when establishing regulatory arrangements will be difficult and when private capital is not readily attracted to take up equity or otherwise participate in a public enterprise. Management contracts also offer a vehicle by which the private sector or foreign private capital can become established in markets otherwise closed to it, as in Cuba.

Obviously, management contracts are most effective when addressing predominantly management-related problems. Their use is most effective when the public owner wants to address such problems while retaining ownership and overall policy direction and when applied to operations that are relatively straightforward and easily duplicated. Where the problems are of a more fundamental nature, more comprehensive institutional arrangements, such as leases or concessions, would normally be required.

Management contracts can be very useful as an intermediate step on the road to other forms of private sector participation. Typically, a government would use a management contract to return the enterprise to a reasonable level of performance in order to attract investors, for example, the contract of the Government of Trinidad and Tobago with two British companies (ECLAC, 1998). Once the water supply and sanitation system is better managed, a long-term concession will be given. The companies holding the existing contract will be given preference in the future granting of a concession.

Mexico City provides a further, but more limited, example. In 1993, the Federal District Water Commission (Comisión de Agua del Distrito Federal) awarded 10-year contracts to four private consortia to renovate and improve the drinking water supply and sewerage services in Mexico City. These contracts, among the largest of their kind, are together potentially worth up to US$ 10 billion. The authorities expect, by this means, to reduce water demand in Mexico City from 35 to 25 cubic metres a second and to have a balanced budget in approximately eight years.

The Federal District has good coverage of services, approximately 97 per cent of households have drinking water supply and about 95 per cent are connected to sewerage. Costs, however, were out of control and between 30 per cent and 50 per cent of water was wasted mainly as a result of deficient billing and a leaky and haphazard delivery system. Metering was limited and many meters were defective or inoperative, with one-third more than 20 years old. As a result, over half the bills were based on estimates, 15 per cent of customers did not receive a bill at all and half of bills issued were not paid. A large number of illegal water connections further compounded the situation.

Very little was known about the asset base, the extent of its disrepair, or the number of connections. Estimates varied on how much it would cost to overhaul the system. According to some estimates, the annual federal subsidy to the Federal District for water and sewerage services was more than US$ 1 billion a year. This is equivalent to the annual sector investment needed to supply the total population of Mexico with adequate water and sanitation services by the end of this century.

To address these problems, Mexican authorities decided to contract the private management for distribution and commercial activities. The two main objectives were to improve cost recovery and to radically improve the water distribution infrastructure.

The Federal District has been divided into four zones of similar size and contracts awarded to a separate contractor for each zone, although, in the short term, the selection of only one contractor could have produced the lowest cost. The division into separate service zones, however, affords opportunities for the implementation of a more effective regulatory incentive structure, based on comparative yardsticks or benchmarks, than that feasible when there is only one contractor. In addition, having several contractors carrying out essentially the same activities in different service zones makes possible a constant revision of the technology used. Moreover, competition could result in more efficient management and it was believed that the system was so large that no single company could complete and manage the system quickly and efficiently (Richard and Triche, 1994). Having several contractors also reduces the risk of service disruption.

Under the present contract, private operators are concerned only with operations and commercial aspects, but not with production. The city retains ownership of the infrastructure and control over the policies, including the implementation of the new billing system. It is a staged contract, so as to allow the private companies to raise the level of information and to gradually assume more responsibility prior to entering into a full incentive contract.

The phased approach will allow the government to control the rhythm of contract implementation through the Federal District and will reduce the financial uncertainty and political risk for both the public authorities and the contractors. The participation of experienced foreign operators ensures that the consortia can draw on their knowledge and experience of operation as well as on state-of-the-art technologies.

Leases The leasing of equipment and machinery, buildings, land, as well as of entire plants, has become common practice in many countries. For a lessee, the main advantage is the avoidance of capital expenditures. For the lessor it is a means of reducing risk as the lessor maintains the ownership of a tangible asset.

In a water utility lease contract, the lessee receives full operational and financial control of the assets essential for the operation of the facilities. Public authorities also use leasing to grant an exclusive right to provide services in a given area or to operate a particular transportation route. This is particularly useful where there is a decision to provide socially desirable but unprofitable services which otherwise would not be attractive to a private entrepreneur (Kessides, 1993).

Since, however the financial situation develops, a lessee must pay rent, a lease contract is a means of transferring the commercial risk related to day-to-day operation and maintenance to the lessee. This transfer of commercial risk offers the advantage of strongly motivating the lessee to improve the efficiency and reduce the costs of operation. Lease contracts, however, must be carefully designed and the public authorities must be capable of fulfilling the essential monitoring and regulatory functions, if the risk is to be actually transferred.

As the lessee rents the facilities and the contract does not transfer ownership, a lessee is not responsible for capital expenditures. A lessee is, however, normally responsible for the financing of working capital, and for the maintenance and repair of the assets in use. Depending on the nature of the assets, there can be shared arrangements, particularly for the replacement of capital components with a short economic life.

The responsibility for fixed investment, such as system, remains with the owner. This makes leases particularly suitable for hydroelectricity generation, drinking water supply, wastewater treatment and similar activities, which require periodic large capital investments, and where responsibility for operation and maintenance can easily be separated from that for major investments. In both France and Spain, this has been an established practice for decades in urban drinking water supply and sewerage (Kessides, 1993). In the United States, there is an increasing interest in leasing, particularly for wastewater facilities (Rogers, 1992).

A lessee assumes responsibility for the regular maintenance of the facilities leased. It is necessary that the condition of the facilities and any needs for rehabilitation are specified in a detailed inventory accompanying the lease. Efficient mechanisms and strong incentives need to be incorporated into the contract to ensure that adequate maintenance will actually be carried out and that a viable asset base will be returned at the end of the lease. Public authorities must, of course, exercise effective control by including in the contract comprehensive specifications for the maintenance required, for the performance criteria to be used for evaluating service quality, for enforcement and dispute resolution procedures and the penalties for non-performance.

A good example of the operation of leasing arrangements is provided by Guinea (Brook Cowan, 1996). In 1989, the Government of Guinea entered into a lease arrangement for private sector operation of water services in the capital city, Conakry, and 16 other towns. The national water authority (SONEG) owns the water supply infrastructure and is responsible for planning and investment, as well as for servicing the sector's debt. It also sets the tariffs. A joint government–private water management company (SEEG) holds a 10 year lease under which it is responsible for operating and maintaining the facilities and billing and collection. There is a separate management contract with the private partner for the management of SEEG.

Despite a lack of clarity of who is responsible for what and the sharing of commercial risk, given the overlap in ownership and contractual relationships, the private partner constructs system expansion for SONEG. Over the first five years the arrangement had made some considerable achievements. The number of connections increased threefold, although one-third were later cut off due to non-payment. Despite this, the proportion of the population with access to safe water increased from 15 per cent to 52 percent. The operations are financially sustainable, with operating costs equal to 71 per cent of operating revenues, but with a tariff of US 90 cents per cubic metre. SONEG has been relatively complacent about tariff increases and this has considerably reduced SEEG's commercial risk and, possibly, incentives for efficiency. There remain, therefore, major problems which the lease arrangement has not solved and which point to the need to define responsibilities very clearly in any leasing arrangement. Managing leases is difficult and not necessarily easier than managing outright privatisation for government authorities.

As the Guinea case illustrates, the separation between major investments and operation and maintenance may distort investment incentives. It makes it easier, however, both to evaluate the lessee's performance and to maintain government control over the expansion of the system (Roth, 1987). The inclusion of the principle of full cost recovery in the rent can help prevent the distortion of investment incentives. When both the lessor and the lessee have to recover costs from the same source, there is a strong incentive for them to co-operate in making sound investment decisions. At the same time, the separation of ownership from the responsibility for system operation and maintenance can be an important advantage in that leasing can yield many of the benefits of privatisation, but usually does not require any special legislation because the public sector retains ownership (World Bank, 1995). Leasing project operation may avoid both the complexities involved in dealing with the financial markets and the need for obtaining legislative approval for increased private participation. Service and management contracts offer the same advantage, but leasing is far more flexible and transfers more responsibility. For example, the lessee usually has complete freedom in the hiring and payment of staff (Vuylsteke, 1988).

The duration of lease contracts will depend on the nature of the facilities involved. They may run from six to ten years, although they can be longer (sometimes up to 30 years). Lease contracts are typically renewable. Long-term lease contracts may evolve into a concession or partial concession or other similar arrangements involving investment, particularly after any costly investments have been completed. Generally, the length of lease contracts should correspond to the amortisation period of the assets created under the responsibility of the lessee.

There are less commonly used variations with distinct market niches, such as lease and lease buy-back arrangements. In leases, the emphasis tends to be placed on the management of an already existing system. Lease buy-back contracts, in contrast, focus more on the building and engineering of new projects. Lease buy-back schemes take advantage of the capacity of the public sector to finance large-scale construction (Israel, 1992). The public sector finances the construction of a facility which is then leased to the private sector for operation and maintenance in return for a fee to cover the amortisation of the initial debt burden. This practice is particularly applicable to projects characterised by elevated up-front costs, high risks to the private sector, and where it is essential to ensure accountability and operational efficiency.

A further variant is known as a sale-leaseback transaction, where a public agency sells a property, say a wastewater treatment plant, to an investor and simultaneously executes an agreement to lease the property back from the buyer under a true lease or a lease-purchase arrangement (Doctor, 1986).

Concessions Concession arrangements are by far the most common form of private participation in water supply and sanitation services, but are less common in other areas of water management. Concessions in the drinking water supply sector are not new; they have been widely used since the 18th century in France and since the 19th century in Spain. In fact, they were common in many parts of the world before the trend in this century to the provision of sanitation services by governments.

Under a concession contract, the owner grants an exclusive right to operate and maintain a facility, whether a whole system or self-contained parts, for a specified period. A concession contract normally transfers the responsibility for financing major investments to the contractor. This implies that all commercial risks and most financial risks are shifted to the contractor. A concession contract may or may not transfer ownership of facilities. In either case, the contractor must return them in good condition at the end of the contract period. The public authority maintains control over service provision through reviewing investment plans and their implementation, monitoring service quality, regulating tariffs, and so on.

Concession contracts are designed to be long enough to allow the concessionaire to recover investment and to provide a reasonable return to the equity investors, typically from 15 to 30 years, and are often renewed. Completely private or mixed public–private companies may hold concessions. A mixed ownership may be preferable in certain areas because it may be more politically acceptable and may reduce the risks for the private partner, but with similar risks to those in the Guinea lease. Where possible, concession contracts with entirely privately owned companies are preferable so as not to dilute accountability and incentives (World Bank, 1994).

A concession contract offers the obvious advantage over lease or management contracts in that it assigns responsibilities for operations, maintenance and investments to a single entity. Where investment decisions are taken in isolation from commercial considerations this can lead to inappropriate investments and wrong technical solutions. The manager of the system is in the best position to forecast demand and make investment and technical decisions that will meet demand in a commercially viable fashion. At the same time, the ownership of the physical assets, even for specified periods, provides a much stronger incentive for adequate maintenance.

For concession projects to be successful, an assured revenue stream is essential to persuade the private sector to commit funds. Governments may find themselves required to enter a binding long-term agreement on the minimum levels of revenues, which it is required to pay. In the case of a drinking water supply project, for example, where the holder of the concession depends on water sales to recover the capital invested, the government may guarantee a minimum level of water sales independent of usage fluctuations. In many cases, the proper use of government guarantees can be an effective and low-cost measure to lower project risks, to turn borderline undertakings into financially viable projects and attract private interest.

Excessive insulation of the private investor from commercial risk, however, can reduce motivation for cost minimisation and create other undesirable incentives. On the other hand, while the private contractors, almost certainly, must seek some guarantees with respect to the availability of project inputs and the possibility of selling its output, they must be prepared to accept some level of risk. The non-exportable nature of project output for projects in the public utilities sector implies that, for foreign investors, repayment and dividends depend on convertibility and transferability of currency and the stability of the foreign exchange market (Traverso, 1994).

The largest single privatisation within the water sector has been the concession of water supply and sanitation provision in Buenos Aires, Argentina. The process and subsequent history of the concession are worth describing in detail as it well illustrates both the strengths and the weaknesses of concession arrangements. Drinking water supply and sewerage services in the Buenos Aires Metropolitan Area at the time of the concession left much to be desired. Forty-five per cent of the population did not have access to drinking water supply and 61 per cent no sewerage. Moreover, 79 per cent of the pipe network had exceeded its useful life, treatment technology was obsolete, only 15 per cent of connections were metered and virtually all sewage was discharged untreated.

Little was known of the basic state of the system. Progressive reduction in funding had had a debilitating effect on the business, both in limiting expansion of the customer base and in maintaining and upgrading the assets. Preventive maintenance regimes had been abandoned, standards had fallen, there was an increasing risk of major breakdown and a lack of replacement and repair of deteriorating assets led to spiralling water losses. Under these conditions sale of the company would have been impossible.

In 1989, the government of Argentina announced its intention to arrange a private concession for drinking water supply and sewerage services in the Buenos Aires Metropolitan Area. The arrangement chosen was a 30-year concession. The selection of the concessionaire followed a two-phase process, technical and financial. The bidding documents for the contract outlined in broad terms the proposed regulatory regimes and the membership of the proposed regulatory agency, but its structure and operational procedures were not detailed in advance. Although there was some uncertainty over the powers of the regulatory agency, the delineation of authority appeared sufficiently clear and the contract was felt to provide an adequate protection to the operator from arbitrary decisions by public authorities. In addition, confidence was expressed that local courts could be counted on to treat the concessionaire fairly. Risks associated with exchange rate devaluation had been removed, at least temporarily, by the government's commitment to sustain full convertibility at a fixed rate.

The process from bid preparation to award took about two years, including a year to prepare bidding documents and draft the legal documents. One of the main problems was the inadequate quality of existing operational and commercial data. For example, revenues could not be audited, little historical data was available about the volume of the water pumped for either the quantity of unaccounted for water or of the quantity of water actually consumed. Consequently, calculated values had to be used. In addition, as maintenance had been inadequate, little was known about the condition of the system and about requirements for rehabilitation and expansion.

The government devoted considerable attention, effort and resources to overcome these shortcomings. All records were made available for bidders to review and personnel were available to answer questions regarding the system. External consultants were engaged to develop background documentation and to promote the concession and identify potential investors. The cost of this preparatory work was about US$ 4 million.

In order to bid, companies were required to demonstrate that they met the requirements relating to technical experience and financial capacity. For the technical phase, service quality and coverage targets were established and bidders were invited to present technical solutions to meet the target goals. They were also required to submit detailed proposals of their intentions over the life of the contract, including an investment plan for the rehabilitation and expansion of the physical system (Richard and Triche, 1994).

The financial selection criterion was the proposed consumer tariff. Since the pre-concession average charge of about US$ 0.40 per cubic metre already covered the operating and maintenance costs of the system and it was assumed that the private operator would be more efficient, bidders were expected to offer initial rates lower than the existing tariff. A baseline tariff was specified and the bidders were asked to submit a price bid expressed as a percentage with reference to the baseline tariff.

In December 1992, the government of Argentina awarded a 30-year concession contract to Aguas Argentinas, a consortium led by the French company Lyonnaise des Eaux Dumez. It won with a bid that would reduce residential user tariffs by about 27 per cent. Under the contract, the company assumed full responsibility for the entire drinking water supply and sewerage system. It had to finance and execute the investments necessary to achieve service targets as specified in the contract, approaching US$ 4 billion.

A regulatory body was established, known as the Tripartite Sanitary Works and Services Authority (ETOSS), with representation from the federal, provincial, and municipal governments. Its mission is to ensure enforcement of the contract, including the conditions of service, investment plans and allowable tariffs. Specific parameters describing water quality standards are set out in the Regulatory Framework Law. The concession contract stipulates tariff rate-making provisions; specifically the use of 14 operating cost components to calculate a cost index by which the tariff can be adjusted. Every five years, the regulator will adjust the index to account for new investments and their effects on operating costs (World Bank, 1995).

In the contract with Aguas Argentinas, the design of the tariff formula is at the core of effective regulation and critical to the sustainability of the reforms. The regulatory regime recognises legitimate costs and allows an additional profit margin (a loose form of cost-plus regulation, as information on cost is very approximate). There are four different kinds of tariff structures. Which structure is applied depends on the kind of building involved, and how the tariff is charged depends on whether the consumer is metered. The design of this family of tariffs introduces a number of distortions. First, the metering incentive is in the wrong direction. Consumers have an incentive to install a meter only if their consumption is small, but without meters it is difficult for the concessionaire to track water losses. Second, average prices are lower for large users than for small users, yet for water, unlike for electricity, there is no technological justification for this. And third, the two-part tariff for metered customers leads to cross-subsidy problems. The tariff has a fixed part to cover the cost of infrastructure and a variable part that is proportional to consumption. Total connection costs, however, are less than total revenue from the fixed part of the tariff. The resulting cross-subsidies lead to inefficient and often inequitable investment decisions, a problem that has worsened since the concession area was increased to cover many suburbs without connections. Another problem with the tariff formula is that it provides little incentive to invest in expanding the sewerage system (Crampes, 1996).

The regulators have shown awareness of the tariff design problems, but they have been bypassed by the holder of the concession who has tended to deal directly with the government. The public policy criteria for testing whether revision is needed should be pre-established and clearly defined. The changes to the contract have been limited to the issues at hand, but there has been a tendency to negotiate outside the regulatory system which has considerably weakened the authority of the regulator (García, 1998).

In general, the relatively short, but complex, experience with the Buenos Aires concession points out the difficulties of controlling very large concession contracts. It would perhaps have been advisable to break up the concession either vertically or horizontally. The ETOSS, the regulator, has been side-tracked, on a number of occasions, in the negotiations of revisions in the contract agreement. In the event, the regulator has been powerless due to the magnitude of the concession and the political importance of water supply and sanitation services in one of the world's largest cities. Finally, the experience points out that no one solution provides an answer to how best to provide service.

A specific variant of concessions is build, operate and transfer contracts (BOT) and its many variations (Table 4.4). In concessions, the emphasis is usually placed on the management and expansion of an already existing system, and a utility company usually leads the concession consortia. BOT contracts, in contrast, focus more on construction and operation of new facilities and BOT consortia are usually headed by a major construction or engineering company.

A BOT contract is a long-term concession to finance, build and operate specific works. At the end of the contract, the company managing the project returns the system, usually at no cost, although the transfer may include a final payment to the equity investors. Typically for a BOT concession, a consortium is formed including major construction and engineering companies and suppliers of heavy equipment, but it may also include a separate management company and, portfolio investors, such as financial institutions.

The advantages to governments of BOT contracts are the control from the beginning of project design, construction, operation and maintenance by the contractor. This control, when combined with the required long-term equity commitment, attracts private sector capital and can be expected to lead to significant cost efficiencies. In addition, the planning, development and management of the project by the contractor may save the utility considerable development, overhead and management costs. Moreover, as the loans are made to the company managing the project and not to the utility owner, BOT arrangements, if properly structured, can circumvent any public debt service restrictions.

BOT contracts are usually for a fixed term, typically 15 to 25 years, but may also have a movable transfer date. Later transfers are normally agreed to if the project has not reached the initial projections of revenues, or a shorter

Table 4.4 BOT and its variants

Variant	Characteristics
AOO (add, own and operate)	A utility retains existing plant. Expansion is financed and owned by the contractor
BBO (buy, build and operate)	A contractor buys assets from a utility, expands capacity and operates them
BLT (build, lease and transfer)	A contractor builds facilities, leases them and transfers them back to a utility
BOMT (build, operate, maintain and transfer)	Equivalent to BOT
BOO (build, own and operate)	Usually through a turnkey contract, a contractor builds, owns and operates infrastructure for a utility
BOOST (build, own, operate, subsidise and transfer), BOOS (build, own, operate and supply) and BOOT (build, own, operate and transfer)	Equivalent to BOT
BTO (build, transfer and operate) and CAO (contract, add and operate)	Used where contractor ownership is not allowed
DBFO (design, build, finance and operate), DBO (develop, build and operate), DBOM (design, build, operate and maintain) and FDBOM (finance, design, build, operate and maintain)	The contractor is accountable within performance specifications. Initially he assumes no commercial risk, assuming it in increments and conditionally as a government sets up appropriate regulations and capital markets
LDO (lease, develop and operate)	Similar to BBO contracts, but the utility retains ownership
RLT (rehabilitate, lease and transfer) and ROT (refurbish, operate and transfer)	A contractor refurbishes plant, brings it on line and transfers it back to the utility

Source: Augenblick and Custer (1990); Israel (1992); Cao (1993); Kessides (1993); WorldBank (1995).

concession period if projections are exceeded earlier. It is important that contracts are of sufficient length so that their amortisation periods correspond closely with the life of the facilities. They can be applied to virtually any activity, including urban and rural drinking water supply, sewerage, wastewater treatment, hydroelectricity generation, provided there is a clear and certain source of revenue.

The negative side of BOT contracts is their need for extensive support and their exceedingly complex nature – legally, technically and financially. They also require carefully developed specifications, particularly for regular maintenance requirements, if they are not to lead to continuous recourse to the courts.

Comparing alternatives

Unlike service and management contracts, concessions and lease contracts have the important advantage that they provide strong incentives to increase efficiency because they transfer maximum commercial risk to the private contractor. Under a lease or concession, as with outright ownership, the holder has to manage the service efficiently in order to earn a profit.

A further advantage of leases and concessions is that, like divestiture, they can be used to exploit the comparative advantage which private entrepreneurs are considered to have over the public sector in identifying new investment opportunities. It is common practice for companies identifying possible new concession projects to be awarded a premium in the subsequent call for bids for operation.

Most lease and concession contracts involve an element of monopoly and, therefore, the private contractor is usually subject to regulatory controls. Private contractors are increasingly inclined to insist that the public authorities should bear a part of the project risk, for example in the form of a guarantee of revenues or asset values. Their insistence is based on the danger that any future changes in regulation involve a degree of risk to private capital (OECD, 1991).

One particular issue that has to be addressed is the fact that many public utilities operate at a loss and will not quickly become profitable. In such cases, sales and concession and lease contracts could generally be bid either on the basis of an explicit, pre-specified subsidy, with the winning bidder quoting the lowest tariff, or on the basis of a given tariff, with the bidder quoting the lowest subsidy (World Bank, 1994). Another possibility would be to transfer control in several phases. In this case, the transfer process should be structured so that long-term financial performance forms the basis for a final valuation, while an interim sale price is based on short-term performance (Hemming and Mansoor, 1988). For example, a government might begin by entering a management or service contract with a private firm, keeping compensation and risk exposure low and, later, transform the contract into a sale, concession or lease.

Lease and concession contracts share a common disadvantage over sale in that they require meticulously developed specifications for the regular maintenance requirements to be provided by the contractor. The final condition in which the facilities revert to the public owner upon completion of the contract must also be carefully and clearly specified. Without detailed specifications, it would be advantageous for a contractor to run down the facility towards the end of the contract period or to design it with a planned obsolescence matching the schedule for transfer. Disincentives for proper maintenance towards the end of a contract can be reduced. One way is to make contracts of sufficient length so that their amortisation periods correspond closely with the life of the facilities and a second is to permit the incumbent to compete for contract renewal.

In this respect, concessions offer some advantages over leases as they provide coincidence between ownership and the responsibility for major new investments with responsibility for system operation and maintenance. The separation of responsibilities, occurring under leases, may distort investment incentives and reduce incentives for maintenance and routine repairs. The integration of the design and construction with the responsibility for operation and maintenance, which characterises concessions and BOT contracts, provides stronger incentives to design the facility to minimise operational and maintenance requirements. It also mitigates against designs and technological solutions, which subsequently lead to costly and unnecessarily complicated operating and maintenance procedures.

Leases, as well as service and management contracts, fail to relieve public investment budgets because they leave the entire burden of infrastructure financing in the public sector, unless the company is self-financing. However, when concession arrangements are not feasible, due to the low borrowing capacity of the private sector or to an unstable political and economic situation, the public sector might have to assume responsibility for investment. In these cases, a lease contract becomes the appropriate commercial arrangement. However, once obstacles to a concession have been removed, it may be possible to convert a lease contract into a concession under which the private company makes limited future investments and pays a rental fee on completed investments (Triche, 1990).

Finally, in many countries, leasing and similar contracts are subject to differences in tax treatment, which can make these arrangements advantageous from the private viewpoint because of the reduction of the effective cost of capital. It should be remembered, however, that whenever cost-benefit analysis from the national accounting stance is attempted, direct transfer payments, debt service, subsidies, taxes, and so on, must be excluded from the calculation, as they represent only a form of income redistribution.

This implies that the effect of preferential tax treatment would be to transfer the burden of meeting finance costs from the project beneficiaries, assuming full cost recovery, to the general tax-payer.

Lack of knowledge about the conditions of the existing asset base is an important impediment to greater private sector participation in provision of public services, particularly in the water supply and sanitation sector. Without detailed knowledge of the condition of the existing asset base, private firms cannot be expected to make a rational decision on whether to participate or not. Nor is it possible for them to conduct meaningful negotiations or formulate a credible financial and technical offer if they decide to participate. The privatisation of drinking water supply and sewerage services in many Latin American cities, such as Buenos Aires, Argentina, Mexico City, Mexico and the attempted privatisation in Caracas, Venezuela faced severe problems due to the lack of good information (Lee and Jouravlev, 1997).

It is where outright divestiture is not possible for political or other reasons that the real potential for franchising through long-term concessions lies. It allows the reaping of most of the benefits commonly associated with private property while avoiding the problems often posed by divestiture. It provides a means to transfer many of the prerogatives and responsibilities of ownership to private operators without actually transferring ownership itself.

Shorter-term franchising arrangements, such as management and service contracts, can be attractive as interim stages towards the development of more comprehensive forms of private sector participation. In addition, other arrangements, particularly BOT (build, operate and transfer) contracts and other similar schemes, can allow the public sector to make good use of the advantages the private sector is considered to have in the identification of investment opportunities.

Franchising is potentially very flexible and can be adapted to virtually any situation. Franchises can range from limited contracts for a specific activity (for example, data processing or equipment maintenance) to comprehensive concession contracts that transfer responsibility for operation, maintenance and even for major investments to the private sector.

Depending on the nature of the arrangement, franchising may be able to capitalise on a number of advantages of the market and privatisation:

1. where the private sector provides financing, the pressure on government budgets can be reduced and economically necessary but politically dispensable water-related infrastructure expenditure can be protected from general budgetary pressures, almost as effectively as with divestiture;
2. when the private sector participates in the operation and management of a

utility (for example service and management contracts and leases), efficiency savings can be realised and scarce public sector managerial capacity freed for areas of the public sector which are not appropriate for privatisation; and finally,
3. due to the possession of scarce specialised expertise and entrepreneurial talent, private companies can provide higher quality services than would be otherwise available to the public sector.

FACTORS TO BE CONSIDERED IN OPENING SERVICES TO THE PRIVATE SECTOR

Experience shows that successful private participation in the provision of water services depends on a combination of factors. These include, in particular, the nature of the services to be privatised and the conditions established in the privatisation process which must seek to promote competition, but at the same time recognise the capabilities and limitations of the private sector. The nature of the formulation of the contract, the subsequent monitoring and regulation of service provision, as well as the financial arrangements, particularly regarding due compensation, the ability to establish a close working relationship between the public officials and the private sector and the maintenance of a fluid and transparent exchange of information are also important considerations.

The nature of the services to be privatised is the single most important condition on which success depends. Contractual relationships are known to be more effective when dealing with observable and verifiable outcomes (Holtram and Kay, 1994). In particular, concessions and other types of franchising are more appropriate with well-defined outputs where there is little technological and market uncertainty and when performance specifications and procedures to monitor them can be clearly described in a contract. The contract should be simple and complete. It should specify pricing structures, services quality levels and so on, for every contingency. Complex or incomplete contracts increase the need for monitoring, enforcement, and renegotiation, that is the need for regulation. The attractiveness of private participation increases where the activity in question is sufficiently similar to existing private sector activities, where there are many potential competitors with the requisite skills, and where sunk costs are low (Kay, 1993).

Since some forms of private participation are more appropriate for less capital-intensive activities, the realisation of its full potential may require functional separation of activities with different capital intensities. One example would be separating the provision of network infrastructure from the supply of services over the network. Private participation may work better where a natural monopoly can be horizontally separated. Where several firms operate under similar conditions and under the same regulatory framework, the regulator has at its disposal multiple sources of information, and this makes it possible to implement a more efficient regulatory framework.

No one arrangement for private participation is clearly superior or is right in all circumstances (Table 4.5). Each alternative has advantages and disadvantages. In part, the advantages will depend on the incentives imbedded in the contract design that make them suitable to particular circumstances. The actual choice of the most appropriate institutional arrangement will depend on its effectiveness, availability and acceptability.

Effectiveness depends on the comparative advantages and disadvantages of individual institutional arrangements *vis-à-vis* each other and on the strength of the managerial and technological skills supporting the particular arrangement.

Availability is another important determinant, because the most effective institutional arrangement may not always be available.

Acceptability is a further important consideration and will depend on a number of factors, including government policy towards private sector participation, the nature of the enterprise, its importance, responsibilities and political visibility and the identity of prospective private sector purchasers or investors. (Hegstad and Newport, 1987).

One reason why a government might find divestiture a more attractive option than franchising could be an anxiety that under franchise contracts it might still have *de facto* some residual liability to fund parts of future capital expenditure (Kinnersley, 1990). Conversely, one reason for choosing concessions over divestiture is fear of default. However, given the strategic importance which society tends to attribute to some aspects of water-related infrastructure, it seems in practice inconceivable that private owners would be allowed to default. For example, it is difficult to imagine that ownership of a large city's drinking water supply and sewerage facilities could be allowed to default and pass into the hands of a receiver. Hence the importance of regulation, for if competition is essentially about fearing loss of market share and monopoly about being free of such fear, the task of the regulator is to maintain the competitive tension.

Public authorities can achieve greater efficiency gains from arrangements to increase private sector participation through competitive bidding processes. Competitive bidding helps to ensure that contracts are awarded to the lowest cost and most efficient providers and helps to guarantee

Table 4.5 Most appropriate use of the alternatives for private sector participation in water services

Type	Water supply and sewerage	Hydroelectricity generation	Irrigation and drainage
Unregulated private ownership	Only likely to be acceptable where there is a large unmet demand	Where the market is large in relation to the minimum efficient scale of generation, an inter-connected system with access on a fair basis, and a developed capital market	Where there are strong local institutions, clearly defined water rights, arrangements for conflict resolution, and liberalisation of the foreign trade regime
Regulated private ownership	With appropriate institutional and regulatory frameworks, a strong regulatory capacity and low regulatory risk	Where some of the above conditions are absent	
Concessions	Where there is no previous experience of private ownership and high regulatory risk		
Leasing	In all sectors where there is an inadequate regulatory framework or political and economic instability		
Management and service contracts	In all sectors this will be generally a transitional arrangement or whenever they are no effective alternatives under any kind of ownership		

Source: Lee and Jouravlev (1997).

satisfactory behaviour on the part of the winner. There must be, of course, a clear and formal separation between the public provider and the private sector bidders. The evaluation of the tenders submitted and the awarding of contracts needs to be transparent and seen to be fair. It must be based, therefore, on uniform and competitive procedures. The control of collusion in any bidding process is obviously of paramount importance.

Awarding a company or contract to the highest bidder merely transfers some of the benefits of any monopoly power the successful bidder may enjoy to the government, but will not protect consumers from future monopoly behaviour (Kay, 1993). In order to maximise future productive and allocative efficiency, activities should be awarded to the bidder who proposes to charge the lowest prices or to offer the best services. The state, by means of contract specification, ensures that the firm maximises allocative efficiency, whereas the firm devotes itself to maximising its profits. This should result in cost minimisation behaviour on the part of the firm.

Since bidders know far more about industry conditions, particularly conditions of technology and demand than the public sector officials organising the bid, it is generally neither feasible nor advisable to specify technological solutions in the contract. Bidding and negotiations should focus more on the cost and quality of service to the customer and not on the details of the technology involved. The contract design itself should provide the incentives for firms to seek out and use the least-cost methods and the most appropriate technological solutions.

In many recent contracts, the so-called 'double envelope' method has been used. At pre-qualification, bidders are required to demonstrate that they meet the requirements relating to technical experience and financial capacity. For the technical phase, service quality and coverage targets are established and bidders are invited to present technical solutions to meet the target goals. This can be useful, for example, to eliminate those bidders who propose risky or untried technological solutions. For the financial phase, the contenders are asked to submit the lowest price bid.

Prior to bidding, whether in sales or concessions, it is advisable for the public authorities to devote considerable attention to establishing the actual condition of the facilities. It is also important to ensure that the information, even if merely fragmentary, is accessible to all potential bidders. Those responsible for the privatisation process should also devote attention to detailing feasibility studies to demonstrate financial viability over the life of the proposed project to potential bidders. In this respect, it is important to remember that ultimately only the certainty of a sustainable adequate earnings stream determines project viability and attractiveness from the point of view of private investors. They should also give attention to analysing the possibility of restructuring the proposed project to promote competition so as to facilitate regulation and monitoring and to make it more attractive to potential investors. Finally, consideration must be given to developing the detailed specifications of the tender documents to ensure that bids are comparable and to designing the contract to ensure that it serves the public interest.

In general, concession contracts should contain comprehensive specifications, including clear, well-defined goals, standards, service standards, and dates for goals to be reached. If these parameters are not clearly specified in the contract, firms cannot make a rational decision on whether to bid or not. Moreover, if they do bid, they cannot provide a credible offer, either technically or financially. The standards and goals should be set in light of the tariff situation, the amount and nature of investments that can realistically be expected, as well as reflect the condition of the existing asset base.

Direct negotiation may be appropriate for some contractual arrangements, for example, management contracts, where non-price considerations, such as technical know-how and experience may override cost considerations (Triche, 1990). The costs of competitive bidding significantly exceed those of negotiating a contract directly, making it less appropriate for smaller projects. Discretionary procedures, however, carry the danger of a lack of transparency.

To achieve greater efficiency gains from franchising, it is even more essential than with divestiture to create effective competition in the tendering process. This must be done to ensure that the most efficient firm wins the auction, to reduce the rent obtained by that firm, as well as to weed out inefficient operators and to curb the potential for corruption. There should also be effective competition whenever the contract comes up for renewal. This can be difficult to achieve. Potential new bidders are aware that the holder of the concession probably has much better information about its actual value than they do, so that they may be reluctant to outbid the incumbent.

One method used to increase competition and to make the private owners or contractors more accountable for the cost and quality of the service is to subdivide larger projects. On the other hand, neither sales nor concessions should be too small, given the high fixed costs associated with bidding, establishing a contract, and carrying out service provision (Richard and Triche, 1994). In particular, consolidation may be required if small projects are to reach international markets, for example regional water supply companies replacing individual municipal ones.

In some areas and for some services the number of potential service providers may be insufficient for competition. For this and other reasons, such as the likelihood of service disruption, it may be desirable to compare private bids against those from the existing public sector (alternatively, government may take steps to foster the development of competitors). Some governments retain a capacity to compete with the private sector, or to provide a residual means of performing essential functions, should the private company fail.

LIMITATIONS ON PRIVATE PARTICIPATION

Despite its many attractive features, private participation in the provision of water services does have limitations. Water management activities in which government control problems are greatest can be especially prone to such difficulties. Most of the problems discussed here are important potential constraints, but many can be resolved or, at least, mitigated through well-specified regulatory or contract design.

There is a danger that the bidding may fail to be competitive. For example, the provision of most water-related services requires specialised expertise, so there may be very few competitors due to the scarcity of requisite skills. The water supply and sanitation concessions made over the last 10 years have been dominated by only three companies. There is always a danger of collusion between bidders, especially if they are few in number and in countries which do not have a history of competitive markets. Bidding assumes a lack of co-operative behaviour between firms as an auction is aimed at extracting a maximum surplus. A natural reaction among companies is to protect against this through collusion. An additional limitation is the fact that when a concession is renewed, the incumbent holder might enjoy strategic advantages that would deter challengers. These advantages arise from the experience gained from operation of the franchise or asymmetries of information on costs and demand conditions in relation to other potential bidders.

In addition, with concessions, problems associated with the observability and transferability of the investment made by the previous holder at the time of asset transfer may distort incentives to invest as well as the competition for the franchise. In general, the valuation of physical assets is more difficult in the water sector than in other sectors. Capital assets in the water sector usually have a longer productive life and a higher component of sunk costs than in most industries, and their valuation is correspondingly complicated. Due to the extreme length of productive life and the often unobservable state of investments, assets are often underground, it can be difficult to estimate their life and appropriate financial value. This can cause and has caused serious disagreements. Although attempts have been made to solve this problem by finding different ways to compensate for sunk investments, the valuation of these investments can create monitoring problems, allows room for discretion and distorts incentives to invest (Bitran and Sáez, 1994). There is an additional problem in the expense of bargaining or arbitration regarding the appropriate valuation, when disagreements arise and this can be considerable.

Given the difficulties of transferring assets, some authors suggest that, in franchises, leases which leave the responsibility for major investments in the hands of the public sector may have advantages over those that transfer this responsibility to the private sector. Such an arrangement can generate even more problems in practice. For example, retaining responsibility for investment in the public sector allows market forces to act only to a limited extent. The separation of investment from operation and maintenance decisions can lead to undesirable losses of co-ordination and distort incentives to invest. It also leaves open the question of how the franchiser determines the level of facilities to be provided. It addition, this approach may have little attraction at a time of over stretched public finance and will confront governments with difficult decisions on how to finance service expansion and how to raise the efficiency of investment.

There are also occasions when there are difficulties of contract specification and administration. If there is technological or market uncertainty in relation to the service in question, then contract specification will be a complex task. In any sector, it would be impossible to cater for every eventuality that might occur in the life of even a medium-term contract, let alone with a permanent sale, and to foresee how they will relate to investments or costs. There will generally be, therefore, a continuing role for the public sector in regulation, and in contract administration. This uncertainty underlines the specific need to include in franchise contracts clauses allowing both parties to renegotiate terms in the event of significant unexpected changes. It also emphasises the importance, for all arrangements involving private participation, of being able to count on a capable and independent judiciary or other mechanisms to arbitrate disputes between the government and private firms.

The difficulties of contract specification in franchising suggest the possibility that short-term contracts may have advantages, because fewer future contingencies then need to be catered for. Longer-term franchise contracts, however, provide opportunity for greater efficiency gains and have other advantages, for example, longer contracts give contractors more time to recover costs and can enable them to increase the scope and quality of service. Even more important is the obvious point that the shorter the term the fewer incentives there are for maintenance thus increasing the risk of mediocre performance (Bouin and Michalet, 1991). Moreover, short-term contracts deter the holder from making investments in sunk assets. It also must be considered that the organisation of frequent auctions involves major costs and all the associated problems will occur with greater frequency. Finally, if the term of franchise contracts is too short, the water sector will frequently be in a state of turmoil.

In concession and similar arrangements, agreements on how to resolve conflicts during renegotiation should be established in advance. Given the inability to contractually cover all contingencies and the fact that franchise contracts in the water sector often run for a decade or more, clauses allowing the parties to renegotiate in the event of unforeseen circumstances should be built into the contract (Guasch and Spiller, 1994). The causes that can give rise to renegotiation, however, should be specified. Common contract clauses simply stating that under certain conditions either party may request a renegotiation of the terms, without criteria as to what events can trigger a renegotiation or how the new terms should be set, may foster rent-seeking behaviour.

An important political barrier to private sector participation arises from the difficulty of establishing a level playing field between public and private companies. The reason for this is that the private contractor's costs are often different from those taken into account by the public sector. Private service providers normally are at a competitive disadvantage with the public sector since they have to recover all their costs as well as pay taxes and make a reasonable profit. Publicly owned utilities, in contrast, often operate at a loss, receive subsidies in the form of grants, subsidised loans, use of public land, staff time and other resources, usually do not pay taxes, and receive abundant assistance in project planning, design, and financial packaging from the external lending institutions. Private participation may be politically impossible unless public utilities are placed first on a non-subsidised, full-cost recovery basis.

Aspects of the broader legal and regulatory environment for public works can also act as significant barriers. For example, accounting laws and practices, laws governing construction contracts, public works laws and conventions may be inappropriate for private sector participation. Similarly, distortions in the overall incentive environment of an economy (the tax regime, import restrictions, labour laws, and banking, foreign exchange and foreign investment restrictions) and the existence of excessive regulations and restrictions on private activities can also inhibit private sector participation. The cumulative effect of regulation is of profound importance. A single restriction or barrier may not constitute a particularly important impediment, but the cumulative effect of many, even indirect, barriers can be such as to seriously deter private investors.

The theoretical arguments and the practical benefits of private ownership of water services are strongest in the tradable goods' industries operating in competitive markets. In these industries, free from substantial market failures, market liberalisation, restructuring and reduced transportation costs can be counted on to supply the beneficial pressures of competition and of contestability. This will reduce the need for the more detailed and intrusive forms of regulation (see Table 4.5). Franchising arrangements provide a means to harness some of the information and incentive advantages of competition to industries that do not operate in competitive markets or have substantial market failures. It can introduce the characteristics and mechanisms of free markets that are associated with efficiency even in natural monopoly situations where direct competition is not possible. Franchising provides a means to institute regulation gradually and it reduces opportunities for regulatory capture and lessens the scope for political interference in the management of water sector utilities.

As the review of the alternatives shows, the means for incorporating private enterprise into the provision of water services are very varied. The most appropriate selection will depend entirely on circumstances. Concessions are more commonly used for water supply and sanitation, but experience shows that they are no easier to regulate than independent private companies.

Not surprisingly, most of the arrangements that are being used in practice to increase private participation in water-based public services are somewhat hybrid in nature. Management contracts incorporate elements of concession, concessions are often in part contracts or leases. There are examples of partial divestiture through the formation of joint public–private companies. There is no valid universal recipe.

None of the alternatives, however, eliminates the need for regulation, the subject of the next chapter, or for ongoing government responsibility for the other aspects of service provision, such as the quality of service, water quality and environmental standards. The public sector must be capable of supervising the private providers of services. Unless entry costs are low, a franchisee or contractor is always in a strong position either to amend the contract or to disregard it. Close monitoring is required to ensure that private providers meet their obligations under whatever alternative for private participation in service provision is adopted.

REFERENCES

Ahmad, Reyáz A. and Moisés Mainster (1995), 'Smooth transitions', *Privatization in Latin America, 1995. A LatinFinance Supplement*, 4–13.

Augenblick, Mark and B. Scott Jr. Custer (1990), *The build, operate, and transfer (BOT) approach to infrastructure projects in developing countries*, Policy, Research, and External Affairs Working Papers Series, 498, Washington, D.C.: The World Bank.

Baumol, William J. and Kyu Sik Lee (1991), 'Contestable markets, trade, and development', *The World Bank Research Observer*, 6 (1), 1–4.

Baumol, William J., John C. Panzar and Robert D. Willig (1982), *Contestable Markets and the Theory of Industry Structure*, New York: Harcourt Brace Jovanovich, Inc.

Bitran, Eduardo and Raúl E. Sáez (1994), 'Privatization and regulation in Chile', in Barry P. Bosworth, Rudiger Dornbusch and Raúl Labán (eds), *The Chilean Economy. Policy Lessons and Challenges*, Washington, D.C.: The Brookings Institution, pp. 329–337.

Bös, Dieter and Wolfgang Peters (1988), 'Privatization, internal control, and internal regulation', *Journal of Public Economics*, 36 (2), 231–258.

Bouin, O. and Ch. A. Michalet (1991), *Rebalancing the Public and Private Sector: Developing Country Experience*, Paris: Organization for Economic Co-operation and Development.

Briscoe, John (1993), 'Incentives key to improving water and sanitation services' (interview with), *Water & Wastewater International*, 8 (2), 28–36.

Brook Cowen, Penelope J. (1996), *The Guinea Water Lease–Five Years On*, World Bank, Public Policy for the Private Sector, Note 78, Washington, D.C.: The World Bank.

Cao, Andrew D. (1993), 'Infrastructure financing methods', *Privatization in Latin America, 1993. A LatinFinance Supplement*, 12–18.

Chéret, Ivan (1994), 'Managing water: the French model', in Ismail Serageldin and Andrew Steer (eds), *Valuing the Environment*, Environmentally Sustainable Development Proceedings Series, 2, Washington, D.C.: The World Bank, 80–92.

Crampes, Claude (1996), *Regulating Water Concessions*, World Bank, Public Policy for the Private Sector, Note 91, Washington, D.C.: The World Bank.

Doctor, Ronald D. (1986), 'Private sector financing for water systems', *American Water Works Association Journal*, 78 (2), 47–48.

Easter, K. William and Robert R. Hearne (1993), *Decentralizing water resource management. Economic incentives, accountability, and assurance*, Policy Research Working Paper, 1219, Washington, D.C.: The World Bank.

ECLAC (United Nations Economic Commission for Latin America and the Caribbean) (1998), *Progress in the privatisation of water-related public services: a country-by-country review for South America*, LC/R.1697/Add.1, Santiago: United Nations Economic Commission for Latin America and the Caribbean.

Galal, Ahmed (1992), 'CHILGENER', paper presented at the World Bank Conference on the Welfare Consequences of Selling Public Enterprises. Case studies from Chile, Malaysia, Mexico and the U.K. Chile. Background. CHILGENER, ENERSIS, Compañía de Teléfonos de Chile, Washington, D.C.: The World Bank.

Galal, Ahmed, Leroy Jones, Pankaj Tandoon and Ingo Vogelsang (1994), *Welfare Consequences of Selling Public Enterprises: An Empirical Analysis*, New York: Oxford University Press,

García, Américo (1998), *La Renegociación del contrato de Aguas Argentinas*, Buenos Aires: unpublished.

Guasch, J. Luis and Pablo T. Spiller (1994), *Regulation and Private Sector Development in Latin America*, Washington, D.C.: The World Bank.

Haarmeyer, David (1994), 'Privatizing infrastructure options for municipal systems', *American Water Works Association Journal*, **86** (3), 43–55.

Hegstad, Sven Olaf and Ian Newport (1987), *Management contracts. Main features and design issues*, World Bank Technical Paper, 65, Industry and Finance Series, Washington, D.C.: The World Bank.

Hemming, Richard and Ali M. Mansoor (1988), *Privatization and Public Enterprises*, Occasional Paper, 56, Washington, D.C.: International Monetary Fund.

Holtram, Gerald and John Kay (1994), 'The assessment: institutions of policy', *Oxford Review of Economic Policy*, **10** (3), 1–16.

Israel, Arturo (1992), *Issues for infrastructure management in the 1990s*, World Bank Discussion Papers, 171, Washington, D.C.: The World Bank.

Izaguirre, Ada (1998), *Private Participation in the Electricity Sector – Recent Trends*, World Bank, Public Policy for the Private Sector, Note 154, Washington, D.C.: The World Bank.

Kay, John (1993), 'Efficiency and private capital in the provision of infrastructure', in *Infrastructure policies for the 1990s*, Paris: Organization for Economic Co-operation and Development, 55–73.

Kessides, Christine (1993), *Institutional options for the provision of infrastructure*, World Bank Discussion Papers, 212, Washington, D.C.: The World Bank.

Kinnersley, David (1990), 'Private water utilities and a new basin agency', paper presented at III Congreso Mundial de Derecho y Administración de Aguas - AIDA III. La gestión de los recursos hídricos en vísperas del siglo XXI, Valencia, Spain: Asociación Internacional de Derechos de Agua (AIDA), Generalitat Valenciana.

Lee, Terence Richard and Andrei Jouravlev (1997), *Private Participation in the Provision of Water Services*, Serie Medio Ambiente y Desarrollo N° 2, Santiago: United Nations Economic Commission for Latin America and the Caribbean.

Liétard, Philippe and Everett J. Santos (1994), 'On the move. Privatizing Latin American infrastructure services', *Privatization in Latin America. 1994, A Supplement to LatinFinance*, 4–12.

Nellis, John and Neil Roger (1994), *Increasing Private Participation*, Session 2 of 3 of the Private Sector Development Seminar, Washington, D.C.: The World Bank.

OECD (Organisation for Economic Co-operation and Development) (1987), *Managing and Financing Urban Services*, Paris: Organisation for Economic Co-operation and Development.

OECD (Organisation for Economic Co-operation and Development) (1991), *Urban Infrastructure: Finance and Management*, Paris: Organisation for Economic Co-operation and Development.

Posner, Richard A. (1975), 'The social cost of monopoly and regulation', *Journal of Political Economy*, **83** (4), 807–827.

Richard, Barbara and Thelma Triche (1994), *Reducing regulatory barriers to private-sector participation in Latin America's water and sanitation services*, Policy Research Working Paper, 1322, Washington, D.C.: The World Bank.

Rogers, Peter (1992), 'Integrated urban water resources management', in *International Conference on Water and the Environment: Development Issues for the 21st Century*, The United Nations Administrative Committee on Coordination, Inter-Secretariat Group for Water Resources, Dublin, 26–31 January 1992.

Roth, Gabriel (1987), *Private provision of public services in developing countries*, EDI Series in Economic Development, New York: The World Bank/Oxford University Press.

Silva, Gisele, Nicola Tynan and Yesim Yilmaz (1998), *Private Participation in the Water and Sewerage Sector–Recent Trends*, World

Bank, Public Policy for the Private Sector, Note 147, Washington, D.C.: The World Bank.

Stevens, Barrie and Wolfgang Michalski (1993), 'Infrastructure in the 1990s: an overview of trends and policy issues', in *Infrastructure policies for the 1990s*, Paris, Organisation for Economic Co-operation and Development, 7–19.

Traverso, Victor (1994), 'The rules of the game', *Project Finance in Latin America, A Supplement to LatinFinance*, 4–12.

Triche, Thelma (1990), *Private participation in the delivery of Guinea's water supply services*, Policy, Research, and External Affairs Working Papers Series, 477, Washington, D.C.: The World Bank.

United Nations Centre on Transnational Corporations (1983), *Management contracts in developing countries: an analysis of their substantive provisions* (ST/CTC/27), New York: United Nations.

Vuylsteke, Charles (1988), *Techniques of privatization of state-owned enterprises. Volume I. Methods and implementation*, World Bank Technical Paper, No. 88, Washington, D.C.: The World Bank.

World Bank (1994), *Mexico. Country economic memorandum. Fostering private sector development in the 1990s. Volume I. Main report*, Report 11823-ME, May, Country Operations Division I, Country Department II, Latin America and the Caribbean Region, Washington, D.C.: The World Bank.

World Bank (1995), *Staff appraisal report. Paraguay. Asunción sewerage project*, Report 13028-PA, Environment and Urban Development Operations Division, Country Department I, Latin America and the Caribbean Regional Office, Washington, D.C.: The World Bank..

World Bank (1997), *Toolkits for Private Participation in Water and Sanitation*, Washington, D.C.: The World Bank.

Yepes, Guillermo (1990), *Management and operational practices of municipal and regional water and sewerage companies in Latin America and the Caribbean*, Report INU 61, Washington, D.C.: The World Bank.

5. Regulation

Taking water management activities out of the public sector has the distinct advantage that it allows market forces to be the major determinant in decisions. It also means that government interventions can be targeted on those areas where market failures are most pronounced. This can be difficult, as effective regulation requires the correct identification of the sources of market failure, and the targeting of regulation specifically on them. The key, however, to appropriate regulatory design is to minimise government intervention. The benefits of the combination of public regulation of private activities decline as intervention and the associated costs increase.

The shift from the reliance on public ownership and bureaucratic control for the provision of water-related services to the reliance on a regulated private firm, which will often have a monopoly, completely changes the demands on the water resource management institutions. It demands a thorough reconsideration of the existing water resource management policies. The privatisation of water-related services forces a major, even revolutionary, readjustment of the role of the state in water resource management. It requires not only that the state withdraw from many activities but that it take on new ones, often of a very different nature, requiring different skills and knowledge on the part of public sector personnel. In water resource management, all the experiences show that privatisation does not just stop with the transfer of assets, but requires continuing adjustments in public policy as the public sector accommodates itself to a regulatory role.

The change in the responsibilities of the public sector with the transfer of direct water management activities to private hands can mean, and has meant, the restructuring of ministerial responsibilities. An example is the transfer of the supervision of drinking water supply and sanitation companies from the Ministry of Health or Public Works to an autonomous regulatory commission. The increasing role given to the market also eliminates many activities within the public sector. The private operators take over responsibilities for many tasks formerly carried out by the government. In addition, planning for the supply of new facilities is left to be determined through competition and the market. Plant operating schedules of power stations will now be decided through competitive bids on the spot market rather than by a central dispatch office. It is the results of these bids which will determine the release schedules for reservoirs, when operations are transferred to privately owned electricity generating companies. The supervision of cultivation plans for irrigation districts is abolished as the individual farmers decide on production. The personnel employed in these activities become private employees. In cases where state companies also had regulatory functions, privatisation necessarily involves the reallocation of these responsibilities to a new and independent regulator.

There are many options available to governments for the regulation of the private sector once it has handed over the major share of responsibility for water management activities. The available variety of regulatory instruments, which are being applied in the countries where reforms have advanced most, to attract and regulate private sector participation in the provision of water-related services will be presented and analysed in this chapter. It is important, however, to stress at the beginning of this discussion that 'effective regulation is necessarily a complex business, and to pretend otherwise is likely to have damaging long-term consequences for the industries concerned. Undue simplification of the initial framework of regulation for privatised monopolies will very frequently lead to the emergence of much more serious difficulties in the longer term' (Vickers and Yarrow, 1988).

Regulation imposes direct and indirect costs on the regulated firms as well as on the rest of the economy, both in terms of money and in terms of resource misallocation (Stigler, 1971). The direct cost of administering the regulatory process includes not only the budget of the regulatory agency, but also the costs borne by the regulated industry. The magnitude of these costs depends on the degree of regulatory supervision and the complexity of the regulatory agency's task. The more rapidly changes occur in the underlying technological and market conditions, the greater the importance of joint costs, and the greater the variety and complexity of goods and services, among other factors, will act to increase these costs.

It is the regulated industry rather than the regulating agency that tends to bear the burden of the costs of regulatory administration (Helm, 1994b). For example, it has been common practice in the United States for regulated firms to invest in large planning units. The responsibility of these units is among other tasks to monitor the conduct of regulators, to attempt to predict future regulatory decisions and changes in regulatory policy, to prepare documentation for regulatory reviews and to support the claims of companies at the time of regulatory reviews. The administrative costs of the regulatory supervision of water companies in the state of New Jersey averaged 0.87 per cent of total revenue for large companies, 5.25 per cent of revenue for small companies and 0.92 per cent of overall revenue (Crew and Kleindorfer, 1985).

In addition to the direct costs of regulation, there are several indirect costs, which are more difficult to quantify. Indirect costs arise where regulation encourages regulated firms:

1. to use transfer pricing, that is, to take revenue out of the regulated part of their business and load costs into it;
2. to expand into unregulated activities;
3. to use the regulated business as a method of subsidising the funding of unregulated activities;
4. to transfer costs to those regulated businesses that enjoy a more liberal arrangement for passing on costs to consumers (Bishop, Kay and Mayer, 1995).

There are also considerable indirect costs incurred in rent-seeking and other behaviour to attempt to outmanoeuvre the strategy of the regulator and to influence regulatory outcomes in their favour. To this one should add the possibility that imperfect regulatory institutions operating with imperfect information and under budget constraint may be unable or unwilling to force prices to their correct levels, introducing new distortions into the economy (Schmalensee, 1974; Jones, 1994). Perhaps the most insidious of these costs are regulation's adverse effects on radical process and product innovation and the tendency to increase production costs and to shield the regulated industry from competition (Schmalensee, 1974, 1995).

THE REGULATION OF MONOPOLIES

In industries where competitive forces are significantly limited, incentives for allocation and productive efficiency depend critically on the regulation regime in which the firms operate. Regulation by the public sector is a response to these problems of market failure. Monopoly regulation is one such response where there is ineffective competition and excessive market power. The aim of monopoly regulation is to correct market failures through either a legal prohibition of the exercise of potential monopoly power or very specific actions to control the consequences. Regulation can include measures such as functional integration and separation, control of pricing and, possibly, control of investment decisions and the quality of the products.

Regulation of monopoly water services allows a government to formalise and institutionalise its commitments to protect both consumers and investors. The goals of any regulatory system will always be diverse, including:

1. the promotion of allocation and productive efficiency;
2. the minimisation of the economic rent obtained from the asymmetry of information between regulator and firm;
3. the avoidance of regulatory capture and;
4. the development of a credible commitment to regulate.

How to reach these objectives simultaneously is the central question for regulatory policy and it is a daunting task, particularly in developing countries.

There are two broad modes of regulation – the regulation of industry structure and the regulation of firm conduct, or behavioural regulation. The regulation of structure establishes the organisations or types of organisations that can engage in different activities. Structure regulation encompasses, for example, controls on mergers, on the market share of incumbent firms, measures for the functional separation of activities through controls on the vertical and horizontal structure in an industry, and the liberalisation of entry. Conduct regulation is concerned with the permitted day-to-day behaviour of organisations within an industry. It includes the regulation of access and product prices, the regulation of non-price anti-competitive behaviour, the regulation of service and product quality and environmental regulation. Conduct regulation exercises direct control over the objectives of the regulated firm, while structure regulation exercises control over the structural environment of the firm, that is it regulates the number and types of firms in the industry, but not their behaviour (Perry, 1984). The regulation of potential monopolies usually requires the use of both modes of regulation.

The regulation of structure is passive while the regulation of conduct is active. The specific content of conduct regulation, however, is largely dictated by the regulations in place for governing industry structure. In order to minimise the scope for failure in government intervention, conduct regulation should, ideally, be reduced to a minimum. Regulators should strive for an industry structure, which provides firms with strong incentives to make socially optimum choices, rather than engage in detailed management of conduct, which can easily become not very different from keeping the industry in the public sector. There is little merit in converting a public monopoly into a heavily conduct-regulated private monopoly. The structure of the industry should be such as to maximise competitive pressure. If structure regulation fails to achieve this then conduct regulation may be hard-pressed to be effective in restraining monopoly power and undue interference in the industry will almost certainly introduce productive and allocation inefficiencies.

Five criteria have been proposed as a basis for judging the effectiveness of regulatory systems (Littlechild, 1983):

1. Protection against monopoly, including both the concern that may exist about the impact of the level of monopoly profits on income distribution, and the economic efficiency concern that the dominant position that a monopoly enjoys in the market may allow it to exploit its monopoly power and become inefficient, causing welfare losses.
2. Encouragement of efficiency and innovation, the incentives that the regulatory system provides for the allocation of resources to research and development, innovation, and technological change.
3. Minimisation of the burden of regulation, since regulation is imperfect and consumes scare resources, it is important to minimise the need for information, to address the problem of asymmetric information, and to reduce vulnerability to regulatory capture.
4. Promotion of competition, because competition, where it can be achieved, is the best regulator. Regulation should seek, therefore, to encourage full and effective competition as a means to the end of achieving economic efficiency and reducing the regulatory burden.
5. Maximisation of both the proceeds from privatisation and the prospects of the privatised firms.

The extent to which structure or conduct regulation should be used is an empirical problem that necessarily depends on industry-specific conditions. Important variables are the ease or otherwise of new entry, the competition afforded by the underlying technological and market conditions and the degree of the asymmetry of information between current and prospective producers.

In water resource management, at one extreme, there is irrigation where there are few questions of natural monopoly due to the ever-present competition between surface and groundwater. In other areas, market liberalisation, restructuring and reducing transportation costs can be counted on to supply the beneficial pressures of competition and of contestability which will autonomously and passively perform a major part of the regulatory function and remove the need for most industry-specific conduct regulation. Structural reform would always be the appropriate policy choice for other water-related activities which do not involve high sunk costs, produce tradable outputs with a wide range of substitutes and can be restructured to ensure effective and undistorted competition. Electric power generation is a case in point, particularly in larger countries, as is confirmed by a large number of successful experiences (Izaguirre, 1998). Appropriate structural reform can help to minimise the need for conduct regulation even in some aspects of water supply and sanitation, for example in bulk supply, wastewater treatment and in rural services.

At the other extreme, until recently there were limitations on potential entry and competition in electricity transmission and distribution, but this is now changing. With current technology, strong limitations on competition remain for drinking water supply distribution and sewage collection. In the operation of drinking water distribution or sewerage collection networks, two or more firms usually cannot profitably coexist in the same area. Even if all barriers to entry were removed, it is very doubtful that new entry would materialise, except at the expense of productive inefficiency due to the prohibitively costly duplication of fixed assets. In industries with such a high degree of natural monopoly, conduct regulation, rather than structural reform and the promotion of competition, is the only appropriate policy response.

In the process of privatisation, structural reform and the promotion of competition is always the best choice if it is a valid alternative. For example, in electricity generation, characterised by fast changes in the underlying technological and market conditions, new entry is easy because change provides the very circumstances in which new entry is feasible. Rapid technological change renders conduct regulation less efficient because rapid change tends to increase the problem of the asymmetry of information between the regulator and the firm. As an industry becomes more competitive, the need for specific industry regulation will decrease. Conversely, industries where the underlying rate of technological change is relatively slow, as in drinking water supply and sewerage, offer the most promising conditions for conduct regulation. Even in these industries, however, a permanently low rate of technological change should not be taken for granted.

The Regulation of Industry Structure

The purpose of structure regulation is to introduce changes in the organisational structure of an industry so that it approximates, as closely as possible, the structure needed for competitive behaviour. The structure of an industry largely determines the conduct of its member firms. Where competition cannot be relied on to ensure socially desirable outcomes then the aim of the regulation of industry structure should be to facilitate conduct regulation. The choice of structure for a newly privatised industry must take into account a number of variables, including the industry's technological characteristics, potential informational asymmetries, co-ordination requirements and transaction costs (see Table 5.1).

Table 5.1 Characteristics of industries in the water sector

Industry	Economies of scale	Sunk costs	Need for co-ordination
Hydroelectricity	Moderate	High	Very high
Irrigation			
terminal system	Low	Low	Low
diversion and distribution systems	Moderate	High	High
groundwater systems	Low	Varies	Low to Moderate
Water Transport			
piers and harbours	Moderate	High	High
port equipment	Low	Moderate	Low
ships	Low	Low	Low
Drinking Water Supply			
piped	High	High	High
non-piped	Low	Low (except wells)	Low
Sanitation			
conventional sewerage	Moderate	High	High
other	Low	Low to Moderate	Low to Moderate

Source: adapted from Kessides (1993).

Inappropriate industrial structure is one of the main causes of regulatory failure. Getting the structure right largely dictates the nature of conduct regulation. Errors committed in delineating structure, such as not isolating elements of the natural monopoly or not creating adequate competition, can considerably complicate conduct regulation and make its task more demanding and its scope broader than necessary.

The opening of markets to competition cannot be achieved by decree. It is difficult to move from a nationalised monopoly to a competitive industry in a single step, even in industries without barriers to competition. In monopolistic industries, the regulator needs the authority and has the duty to ensure the success of the transition process and to remove, as far as possible, the obstacles to competition without disrupting supply. The mere repeal of the statutory monopoly is not always sufficient to ensure effective competition, especially if the holder of the monopoly retains significant market power and has at its disposal a range of instruments of strategic entry deterrence or of exit inducement.

As a matter of general principle, regulatory policy should seek to isolate the natural monopoly elements in an industry. The rules should prevent firms entrusted with activities with natural monopoly characteristics from extending their monopoly powers. One alternative is to detach, by means of restructuring or contractual arrangements, the potentially competitive activities from those that are natural monopolies or characterised by other categories of market failure and opening them to various forms of competitive provision. It may be more effective, however, simply to rule that within the company the monopoly and non-monopoly aspects of their affairs be accounted for separately.

When sunk costs are not pervasive, but rather are centred in a particular sector of an industry's operations, their damaging consequences can be quarantined (Baumol, Panzar and Willig, 1982). Many elements of market failure in the provision of water-related services are associated with some very specific elements of market structure and do not occur in all parts of a water management system. The appropriate policy response to the non-monopoly activities is the application of general competition and anti-trust policy.

It is common in water-related services for the providing company to take the form of an integrated utility responsible for all aspects of service provision. This, however, may not be the optimum arrangement and functional separation of activities can considerably increase market and competitive pressure and reduce the burden of the regulator (Figure 5.1).

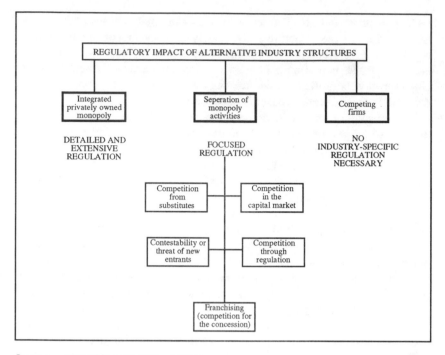

Source: adapted from World Bank (1994.

Figure 5.1 Functional separation increases the options for competition and facilitates regulation

The problem with the integrated model is that the firm may gain an unfair advantage over other companies in the competitive areas of its activities. Functional separation promotes competition by separating out the potentially competitive elements. The objective is to leave the maximum number of operations to the free market and to limit industry-specific conduct regulation to the monopoly segments of the industry.

The separation of functions will generally increase the spectrum of opportunities for private sector participation by, for example, increasing the scope for franchising, and reduce the asymmetry of information between regulator and firm through improving the quality and quantity of the information available to the regulator. It should also help to improve management accountability and increase efficiency by allowing firms to specialise.

The potential benefits of functional separation, however, are counter-balanced by concerns such as the encouragement of a higher rate of service expansion and the achievement of higher service quality. To decide whether and how to effect separation, factors such as the technical, technological and economic constraints to separation as well as the specific legacy of the history of the institutions need to be carefully analysed. Attempts to separate closely interdependent activities can impose high costs, including the loss of the economies of scale and scope for each firm as its size and service area fall. The costs of sector restructuring always needs to be carefully weighed against the potential benefits of cost-minimising behaviour under competitive pressure.

The network characteristics of many water sector activities raises the possibility that the efficient operation of the system as a whole will not be achieved without adequate central co-ordination. Its absence can impose high transaction costs as the co-ordination over a network among several independent and rival firms is inherently more difficult and costly and less effective than within a single organisation. The experience of successful integrated organisations questions the wisdom of excessively dismembering integrated production, particularly where integration involves significant technological and transactional economies, and suggests that undue fragmentation can lead to serious misallocation of resources and over-investment. Fragmentation of responsibilities for planning, investment, operations, maintenance, and debt services may lead to lack of accountability and inefficiency because the firms will not have an appropriate level of control over decisions and actions that affect their efficiency.

At the same time, the existence of potential benefits is not sufficient to ensure that they will be effectively realised under monopoly provision. It is possible that the inefficiencies resulting from the sacrifice of economies of scale would be more than offset by efficiencies resulting from a competitive market structure and advantages of flexibility and proximity to clients that smaller organisations usually enjoy. Very large organisations suffer from organisational diseconomies of scale. In the real world, where a monopoly without external pressure usually undertakes only limited cost-minimisation activity, the introduction of competition, even in a naturally monopolistic industry, can result in lower costs.

In the long run, in reality, this dilemma may not be serious. The costs of making wrong separation decisions in industrial structure may not be very large (Vickers and Yarrow, 1988). In theory, at least, the subsequent evolution of market structure through mergers, new entry, division, take-overs and so on will take care of the problem. This suggests that the best course is to try to maximise competition at the moment of privatisation. Unfortunately, as has already been discussed, the underlying technological and market conditions of many water-related activities are such that there are limits to substantially increasing competition.

Any restructuring, however, is much more easily done while an industry is still state-owned. This is because, once an industry with characteristics of a natural monopoly has been transferred to the private sector, its most valuable asset may not be the physical infrastructure, but rather the licence or right to provide the monopoly service under specified conditions (Guislain, 1992).

The form of an industry and the regulation to which it is subject has a direct impact on the valuation of the assets to be privatised. Uncertainty with regard to the applicable regulatory regime and future restructuring is bound to result in lower investor interest and might attract only those entrepreneurs who have greater lobbying power or a greater willingness to take risks. It is also likely to reduce incentives to invest. In addition, subsequent restructuring might constitute a breach of faith and uncertainty about future reforms might seriously undermine investment planning and increase the cost of capital (Vickers, 1991).

These considerations imply that the success of privatisation can depend on resolving such issues as functional separation and establishing a transparent and credible regulatory framework prior to sale. These are necessary with a view not only to achieving managerial efficiency, but, more importantly, to preventing the consolidation of private monopolies in competitive markets (Bitran and Sáez, 1994). If these issues are postponed, uncertainties about their resolution will make it difficult to promote competition once market structures are consolidated and property rights allocated.

Horizontal restructuring

Horizontal restructuring means separating companies by markets, by geographical regions, by service categories or by individual units. It is an attempt to create a competitive market structure, although local monopolies may persist. A typical example of horizontal restructuring is the limitation placed on the proportion of an industry that can be held by one owner.

Ideally, horizontal restructuring will lead to direct product competition, as it does in hydroelectric generation. If this is the result, the need for conduct regulation dissipates. Increased competition will also improve the efficiency of any state-owned enterprises that may remain in the sector. Even the existence of only limited competition is desirable because it both reduces the need for conduct regulation and enhances its effectiveness by improving the information available to regulators. The existence of a number of regional monopolies can constrain monopolistic behaviour, encourage new entry, promote market contestability, and impede collusion.

Even in drinking water supply and sewerage services, the existence of horizontally separate companies provides opportunities for direct competition for larger industrial and commercial customers. The duplication of water mains or sewers almost always implies a loss of economies of scale, but direct competition for larger customers can be beneficial, especially if there is some product differentiation, such as different qualities of water or types of wastewater treatment.

The existence of regionally defined drinking water supply and sewerage utilities allows competition between contiguous utilities for the right to supply new customers at the boundaries of the service areas. The greater the number of utilities, the greater the scope for such competition, but the greater will be the losses of economies of scale for each utility as its size falls.

For example, both in England and Wales and in Chile, spatial competition currently is encouraged for supplying new developments outside existing allocated company areas. In England and Wales there is also cross-boundary competition where a company must respond to requests for water from any customer regardless of location.

Horizontal separation in the electric power generation sector encourages inter-utility competition, a conclusion that is confirmed by the generally successful experiences in many countries. In Argentina, to cite one example, the three major former federal government-owned utilities were replaced by more than 20 private generating companies and there is an intensively competitive spot market for power (Torres, 1995).

On the other hand, if effective competition cannot be introduced through horizontal separation, on balance it may be better to maintain integration. The effects of horizontal separation would probably be negative in smaller economies with small power systems because of the inability to exhaust economies of scale at the level of individual generators. In smaller and less developed countries, the market may be too small to support enough firms to achieve truly competitive behaviour, except at an unacceptable loss of economies of scale.

Effective competition requires the existence of a sufficient number of firms to avoid collusion and gaming. In the restructured industry, firms should be similar in size and cost structure, adequate transmission capacity should exist and transmission costs should be low. Entry into the industry, through the obtaining licenses or constructing new plant, for example, should be easy and rapid and incumbents should not enjoy important cost advantages unavailable to new entrants (Bacon, 1994). Horizontal separation along regional lines may tend, however, to encourage companies to develop geographic market sharing. As a result, competition would be constrained through tacit collusion not to compete for each other's markets.

However, even when horizontal separation leads to local monopolies, unless there is no correlation in the cost conditions among them, it enables regulators to have access to information from a group of independent providers of comparable services. Such comparisons across those firms can be very useful for setting incentives, based on relative performance benchmarks, and hence opportunities for the implementation of more effective regulatory incentive structures than those that are feasible when there is only one firm.

The informational advantages of the existence of many similar firms are likely to outweigh the loss of economies of scale or scope where a regulated industry is mainly an aggregate of several local monopolies, as in drinking water supply and sewerage. These benefits will increase with the greater similarities in the environment in which the firms operate. The existence of a greater number of similar firms assists regulators by providing them with information from various sources. It also implies, however, that regulatory agencies will be faced with the prospect of regulating and monitoring a more complex industry with different tariff structures and service quality standards, as well as with variations in costs and other conditions. Obviously, this poses a serious challenge to any regulator.

Horizontal division of an industry regionally provides the possibility of competition in the capital market. Horizontal separation regionally also facilitates acquisition and reorganisation of the poorly performing utilities as well as the generation of comparative information with which shareholders can assess performance (Bishop and Kay, 1989). It may be necessary, in order to maintain such a structure to place restrictions on the concentration of ownership and many countries have such rules. The large size of public utilities, some of them will rank among the largest companies in any economy, can act as a barrier, in itself, to concentration of ownership.

For example, concerns over the concentration of ownership have led in England and Wales to rules calling for separate Stock Exchange listings for all regulated water companies after take-overs, mergers or where a company outside the water industry wishes to acquire a regulated water business. (Murray, 1995). The regulator, following the merger between Lyonnaise des Eaux and Northumbrian Water, secured agreement from Lyonnaise des Eaux to separately list its water interests in England and Wales on the Stock Exchange by the end of 2005 (OFWAT, 1995b and 1995c). In Chile, the drinking water supply companies are classified as large, medium or small and no one will be permitted to own more than one large or two medium-sized companies.

Capital markets also act as a powerful disciplinary force for poorly performing regulators. The market valuations can change following regulatory actions. For example, comparing the returns for a regulated company or industry with the returns for a comparable sample of unregulated companies provides a useful way to test whether there is regulatory capture (Dnes, 1995). Capital market response also provides feedback to the regulator, influencing its decisions. It can also constrain regulatory discretion through the performance expectations of shareholders and customers. The role of the capital market is particularly important because of the issue of government or regulatory failure and, even in the absence of such failures, regulators may be ill-informed about the financial consequences of their decisions for the regulated industry.

Vertical restructuring
Vertical restructuring separates activities previously performed by an integrated vertical monopoly. A typical example is the division of a state-owned power utility into separate generating, transmission and distribution companies. Three main factors favour vertical integration: technological economies, transactional economies, and market imperfections such as imperfect competition, externalities and imperfect or asymmetric information (Perry, 1989). Vertical integration in unregulated firms subject to a reasonable degree of competitive pressure generally promotes efficiency and increases welfare. The negative consequences of a vertically integrated industry structure are felt in the absence of competition. Even in the absence of competition vertical integration can usually be justified on efficiency grounds. Unfortunately, in regulated industries with natural monopoly elements, vertical integration tends to be associated with lower economic efficiency.

For example, vertical integration can allow a natural monopoly to extend its monopoly power, particularly in network services. It can discriminate in its own favour or in favour of affiliated firms, increasing barriers to entry and foreclosing competitors by means of prohibitive network access charges or discrimination in other terms of interconnection, such as the quality of access. The existence of regulation in one of the market segments can greatly enhance the incentive for a firm to use its market position to extract profits from non-regulated segments (Yarrow, 1991). Ownership of network facilities by a vertically integrated firm, however, is not necessarily a decisive obstacle to the emergence of competition. Advances in technology, changes in factor prices and in other market conditions can erode the advantages of vertical integration and create opportunities for new entry and competition, as is occurring in electricity and gas distribution.

However, vertical integration does tend to hamper conduct regulation and, in practice, it can be difficult for the regulator to hold in check anti-competitive behaviour from vertically integrated firms. Vertical integration usually worsens the asymmetry of information between the regulator and the company as it provides many opportunities for circumvention of information requests. In consequence, to be effective, conduct regulation has to be more proactive and intrusive.

Any potential benefits of vertical separation have to be balanced against the loss of economies of scope and scale, the costs of sector restructuring, and the possible loss of some internalisation of externalities. If economies of scope are significant, there may be a case for the continuation of a vertically integrated monopoly. Even if parts of an industry must remain vertically integrated, conduct regulation or measures of partial vertical separation will be needed to establish control over firm behaviour.

There are various options available for countering the negative effects of a vertically integrated monopoly. The alternatives include using antitrust law to limit anti-competitive behaviour, publishing the terms of negotiated agreements, imposing terms if the parties to any dispute fail to agree, and imposing public service obligations for interconnecting firms (Guasch and Spiller, 1994).

In the water-related public utility industries, it is much easier to restructure the industry vertically in electricity than in drinking water supply. It is true that, in the latter, there are two potentially separate activities – water distribution and sewerage collection. Increasing competition through vertical structural reform is, however, extremely limited because of the strength of the natural monopoly conditions, which derive from established water and sewer networks. There are three main obstacles to structural reform. The first is the need for extremely tight co-ordination between the two services, due to the interrelated demand. The second is the high cost of service delivery, in relation to the cost of water production or wastewater treatment, and finally, the fact that the experience gained and the equipment used in one is useful for the other. Nevertheless, there are examples of vertical separation of activities, which perform well in many countries.

In the electricity sector, natural monopolies have existed traditionally in transmission and distribution. There is also, however, a need for very close co-ordination between generation and transmission, since demand fluctuates randomly, supply is subject to unpredictable outages and equilibrium must be maintained continuously throughout the system (Armstrong, Cowan and Vickers, 1994). In the past, this has provided powerful arguments in favour of a policy of vertically integrated monopoly for generation and transmission and explains why in most countries the two activities have typically been so integrated. Other arguments in favour of integration include optimal investment and capacity planning and operational co-ordination (IEA, 1994). It is also claimed that integration may also facilitate the handling of power disruptions and supply emergencies.

The problem with the integrated model has always been that control over the transmission network gives the owner an enhanced ability to deter new entry and to discriminate in favour of its own subsidiaries (Paredes, 1995). The separation of generation, transmission and distribution creates conditions for effective competition and encourages new entry. Various forms of competition between generators become possible under vertical separation, ranging from contract competition to supplying the transmission grid under long-term contracts, which may be tradable or not, to spot market competition.

Spot markets can make prices volatile and unpredictable so that contracts between generators and large customers are widely used. Competitive discipline is maintained through contract competition. Long-term contracts offer generators adequate insurance against risks, but contract specification is complex and inefficiencies can arise because every eventuality cannot be covered. Recent technological developments have much improved the possibilities of competition in all stages of electricity distribution and transmission so that many of the arguments for vertical integration no longer hold.

In countries where the power market is small in relation to the minimum efficient scale of generation, suitable sites for new generating plants are few, transmission capacity is insufficient or the costs of sector restructuring are very high, the potential benefits from vertical separation will be limited. This is because any efficiency gains from increased competition would be offset by the loss of economies of scale and scope and the additional co-ordination costs.

Nevertheless, an energy system that includes independent power producers has a number of important advantages over the traditional integrated utility. In addition to the benefits from competition, through having a number of independent generators, independent generators also have powerful incentives to ensure reliability and operate plants at optimal standards.

The diversification of activities is a common tendency of firms in many regulated industries. There is an obvious incentive for any regulated company to generate earnings from activities which fall outside the controls of regulatory authorities and the stronger the regulation of the core regulated business, the stronger the incentives are to diversify (Freeman, 1991).

Many of the private British water companies have diversified into non-regulated businesses ranging from operating utilities overseas, to providing environmental services, waste management, engineering and environmental consulting and process engineering to computer information technology, bottling mineral water and television (Nakamoto, 1991). Similarly, in France, the major water companies have diversified overseas and by taking over other urban services such as sewerage and sewage treatment, solid waste, public transportation, heating networks, cable television and even funerals, or absorbing or being absorbed by public works companies or engineering consultants (Barraqué, 1993). Chilean electricity generating and distributing companies have invested widely in other Latin American countries as well as in the drinking water supply and sanitation sector.

Diversification is generally accepted to be beneficial and desirable, as it allows a company to spread risks and compensate for fluctuations in demand. It can also improve management and productive capacity utilisation and provide protection from concentration in declining markets. Diversification of regulated firms or a merger between two regulated firms, however, can be of concern to regulators. It can expose the utility to the risk of failure from difficulties in its non-regulated business. Diversification into riskier areas can increase the cost of capital, which may influence tariff setting, and the operation of non-core businesses might consume excessive amounts of management time and resources.

This means that diversification of a regulated firm or mergers between regulated firms can worsen the asymmetry of information between regulator and firm. It can reduce the regulator's ability to implement benchmark competition either by reducing the number of available comparable companies or by affecting the comparability of the firm with others.

Diversification can also lead to cross-subsidisation through transfer-pricing in intra-company transactions, which could result where the regulated business pays higher prices than the market rate for the goods and services provided by associated companies. Transfer-pricing can be used to circumvent economic regulation and to support anti-competitive behaviour of affiliated companies. To cope with this problem, a regulator would need to impose an obligation to use competitive tendering or other appropriate methods of market testing to ensure that the work contracted out to affiliated companies is being done at competitive rates. To prevent cross-subsidisation, it is also important to ensure that there is an appropriate allocation of costs between the regulated and the non-regulated business.

However, rules for cost allocation can be difficult to determine, given the asymmetry of information (Armstrong, Cowan and Vickers, 1994).

For these reasons, regulatory policy should be prepared to impose some constraints on the diversification of regulated firms into unregulated activities (Rees and Vickers, 1995). An important question for policy is how to structure the regulatory system to take advantage of diversification while avoiding the undesirable effects. The possible measures available to regulators include requiring separate accounts and the issuing of separate financial statements for the regulated and unregulated activities and imposing an arm's length relationship between the two parts of the business (IEA, 1994). Other possible controls are the close supervision of dividends, loans, asset transfers and other financial transactions between the regulated business and the other activities and prohibiting the regulated business to lend, extend guarantees, pay dividend or transfer assets to the other companies without the regulator's consent.

In England and Wales, the directors of drinking water supply and sewerage companies are obliged to ensure that there are sufficient financial and managerial resources to run the core business, and that they certify that this remains so after a diversification (Armstrong, Cowan and Vickers, 1994). In addition, water companies are under a legal requirement to conduct their business at arm's length from other companies within their group structures and ensure that there is no cross-subsidisation between the regulated business and non-regulated activities or any associated company. The controls over inter-company transactions have been strengthened through both formal licence amendments and the issuing of guidelines and reporting requirements for the companies.

The form of price regulation has important implications for the incentives of the regulated firm to diversify. Regulation of prices based on the rate of return on capital method might give the firm an incentive to expand into other regulated markets, even if these operate at a loss (Averch and Johnson, 1962). This can happen where expanding into other markets enables the firm to increase its rate base and hence permits it to earn a greater total profit than would have been possible without such diversification. This behaviour may discourage new entry and drive out other, even lower-cost competitors. Rate-of-return regulation of tariffs can give a firm incentives to misreport cost allocation reporting any expense as attributable to the regulated service or, failing that, as a common cost. It may also choose to charge prices below marginal cost in any competitive market included in the set of core activities subject to an aggregate rate-of-return constraint (Breautigam and Panzar, 1989).

Regulation of tariffs based on a maximum price limit, price-cap regulation, at least in principle, should induce a firm to diversify into a competitive market only if diversification is efficient. This should greatly reduce any incentives to misreport cost allocations or choose inefficient technology. In practice, however, given the asymmetry of information and the fact that price-cap regulation usually incorporates some of the features of rate-of-return regulation, the same problems are still of concern (Armstrong, Cowan and Vickers, 1994).

THE REGULATION OF CONDUCT

The purpose behind regulating a firm's conduct is to try to ensure that its behaviour is in the public interest. The regulation imposes constraints on the actions of the regulated firms. It can include regulation of the prices charged for the product or service, of the price for access to a network, of non-price behaviour such as a policy against anti-competitive behaviour, the control of service and product quality, of quantity, of investment levels and environmental regulation. The key issue in conduct regulation is how to regulate all the relevant aspects of a firm's conduct simultaneously and this is why emphasis is put on the regulation of prices or tariffs.

The purpose of regulating prices is, in the first instance, to protect consumers from exploitation by monopoly providers. However, it is also a fundamental tool for the creation of a macroeconomic and regulatory environment to provide sufficient incentive to firms to invest and operate efficiently. The major concern in price regulation is to establish incentives for the regulated private firm to act so as to maximise social welfare. Regulation must achieve such behaviour in a situation where the interests of the firm and society usually diverge and where the information available to the regulator and the firm is asymmetrical in favour of the firm (Lee and Jouravlev, 1997).

Two main mechanisms of price regulation have been developed. These are commonly known as the rate-of-return system and the price-cap system. The two systems, although they are often contrasted in the literature, really fall along a continuum. At one end lies the fixing of prices based on actual costs plus a permitted rate of return on the capital invested, and at the other, determining prices by fixed fee or price-cap based in large part on the rate of return on capital. Under the rate-of-return system of price regulation, a firm receives little or no incentive through the tariff to be more efficient. In contrast, under the price-cap system, a firm has full incentive to reduce costs, as it keeps all the benefits from cost minimisation behaviour until the next review of prices (see Figure 5.2).

The choice of a specific price setting mechanism will depend upon the characteristics of the regulatory environment and particularly the underlying technological and market conditions. Much will depend on market structure

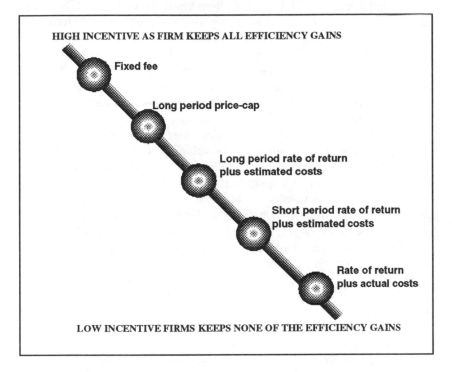

HIGH INCENTIVE AS FIRM KEEPS ALL EFFICIENCY GAINS

Fixed fee

Long period price-cap

Long period rate of return
plus estimated costs

Short period rate of return
plus estimated costs

Rate of return
plus actual costs

LOW INCENTIVE FIRMS KEEPS NONE OF THE EFFICIENCY GAINS

Source: adapted from Jones (1994).

Figure 5.2 Incentives under different pricing mechanisms

and a country's institutional history. Other relevant factors to take into account in choosing a price regulation mechanism include the costs of monitoring company performance, the extent of the information asymmetry between regulator and firm, the ability to design and implement the chosen regime and the scope for efficiency improvements in the industry.

The Rate-of-Return System of Price Regulation

The regulation of price through a rate-of-return system, sometimes known as the cost-plus or cost of service mechanism, has been the most common method used in many countries, including the United States. It ensures that the regulated firm earns a 'fair' rate of return on its invested capital. It scores well on the criterion of restraining monopoly power, but poorly in terms of securing maximum operating efficiency, as it puts emphasis on allocation efficiency rather than productive efficiency (Kay, 1993).

The *ex post* version of rate-of-return regulation, in which the firm is reimbursed for its actual costs, including the opportunity cost of capital, provides no incentives whatsoever for cost reduction. The version used in the United States (or ex ante), where the last period's costs serve as the basis for the current period's prices, provides more incentives. A series of empirical studies of the United States drinking water supply industry, however, has failed to find significant differences in the relative efficiency of privately owned utilities subject to rate-of-return regulation and publicly owned utilities. These results would seem to confirm the low incentive properties of rate-of-return regulation (see Feigenbaum and Teeples, 1983; Byrnes, Grosskopf and Hayes, 1986; Lambert, Dichev and Raffiee, 1993).

Under the rate-of return system, tariffs are calculated to give a 'fair' rate of return so that total revenues equal total costs, including the cost of capital. For example, in the electric power industry in the United States prices have been set to reflect the historical costs of providing supply to each class of consumer. Rate increases are based on a utility's revenue requirements and are set to provide a reasonable return on equity. More recently, the regulatory agencies have required utilities to consider marginal cost pricing, and several states are now calculating prices based on long-run incremental costs and are implementing more complex rate schedules including time-of-day and seasonal rates.

Prices have to be reviewed frequently, often annually, if utilities are to earn an acceptable rate of return each year. The rate-of-return system provides, therefore, better insurance to regulated utilities against cost movements than price-cap regulation will with its less frequent price reviews. In the United States, for example, suppliers can apply for a rate review at any time. Filing a rate case and obtaining governmental approval to change prices tends, however, to be time-consuming and expensive. In other countries, however, rate-of-return regulation does not provide for this same flexibility.

In the United States regulators have authority to require disclosure of financial information and to direct how accounts will be maintained. They can examine the reasonableness of investments and exclude them from the rate base when such investments are considered imprudent. The regulatory bodies have independent and professional staffs who are protected by law from dismissal for political reasons. Open hearings and transparent information on costs and performance help ensure the integrity of utilities. Customers, investors, suppliers, environmentalists, and the public are all able to participate in the regulatory process.

The system, as it has developed, however, has certain weaknesses. The costs of regulation tend to be high, especially for parties participating in studies and hearings, and delays can be frequent because of the quasi-juridical proceedings for rate changes. Rate-of-return regulations encourage utilities to seek rate increases to cover increases in costs rather than reduce costs or increase efficiency. Utilities that over-invest and maintain excess capacity usually are able to pass on these costs to consumers. Moreover, apart from regulatory directives, there are no cost or market-based incentives to adopt least-cost investment, load management, or energy conservation (World Bank, 1992).

In recent years traditional rate-of-return regulation has been criticised because it is difficult to define a 'fair' rate of return, and because:

1. it provides poor incentives to minimise costs and innovate;
2. it encourages the firm to use an inefficiently high capital/labour ratio for its level of output;
3. it requires too much detailed knowledge of the industry on the part of the regulator.

The fact that rate-of-return regulation is based on capping profits rather than prices means that poor incentives are provided for cost minimisation, except in a limited way. Since prices are set so that the regulated firm is assured of a specific return on its investment after recovering its costs and its prices are reduced in step with decreases in costs, it may have relatively little incentive to engage in cost reduction. A particular problem is an incentive to employ too much capital, often called 'gold-plating' or the Averch–Johnson effect (Averch and Johnson, 1962). This occurs because, although there are restrictions on the return that the regulated firm may make per dollar of capital, there are no direct limits on absolute profits, providing a subtle incentive to expand capital stock to increase the total return (Boadway and Wildasin, 1984). Because of this problem, regulators are forced to scrutinise expenditures carefully. Regulators usually require approval for new investments and exclude excess capacity from the definition of the rate base.

Conversely, to limit underestimates of demand, regulators have authority to penalise a firm if rationing occurs (Lewis and Sappington, 1988).

The need for detailed regulation and the perceived insufficient incentives under rate-of-return regulation have led United States regulators to move to versions of rate-of-return pricing which are closer to the price-cap system developed in the United Kingdom.

The Price-cap System of Price Regulation

The price-cap system of price setting attempts to avoid the problems associated with the rate-of-return system, particularly its tendency to put upward pressure on costs, and seeks both to limit the scope for regulatory failure and to reduce the burden of regulation. Three central ideas have been influential in its design:

1. regulation should be based on the control of prices rather than profits;
2. discrete periods between regulatory reviews should be substituted for continuous intervention;
3. regulatory 'contracts' should be established.

The regulation of prices rather than profits provides stronger incentives to improve efficiency and to innovate in production technology and service offerings. It can also help promote competition and aids in concentrating regulation precisely on the particular services where monopoly power is greatest, thus ensuring that consumers are effectively protected against abuse. Price-cap regulation can be seen as a form of contracting where 'the state, through the regulatory licence, contracts certain outputs and services to the private sector, in return for which the private sector is entitled to a revenue stream which is linked to inflation' (Helm, 1993).

The price-cap system was developed in England and Wales where the Office of Water Services (OFWAT) is responsible for the economic regulation of the privately operated water supply and sewerage companies. The primary duty of OFWAT is to secure that the companies properly carry out their functions and that they receive reasonable returns on their capital (OFWAT, 1995a).

The companies must restrict the rate of growth in their prices by the retail prices index adjusted by a factor known as the K factor, which may be positive or negative. The adjustment varies among companies and through time, and is designed to allow companies to finance their investment programmes, while encouraging them to be efficient. The K factor is fixed for 10 years, although either OFWAT or a company can apply for a review after five years. To cover the industry for unforeseen cost increases, reasonable extra costs can be passed through to consumers outside a formal periodic review. OFWAT can use the same procedure to reduce prices under some circumstances. The objective of OFWAT, as of all regulators using this mechanism, is to set the price-cap so that an efficient company has just sufficient income to finance itself.

The procedure used takes various factors into account, including company asset valuation, capital investment programmes, comparative efficiency of company operations, legislative requirements balanced, where possible, by customer requirements, and a fair return to shareholders. Efficiency targets are set in relation to both operation and capital expenditures and take into account comparative levels of service. The relatively long period over which prices are set provides a strong incentive to the companies to improve efficiency.

In the event, OFWAT has found the principle of setting prices for long periods difficult to sustain. It has intervened frequently to alter prices. For example, in 1991, only two years after privatisation, OFWAT wrote to several companies suggesting that they limit their price increases on account of their unexpectedly high profits. The companies chose to follow the regulator's suggestion and a year later, in 1992, prices were lowered for 17 companies, as construction costs had fallen below the levels assumed in 1989.

In fact, OFWAT's actions have had much in common with rate-of-return regulation and in practice its form of price-cap regulation has turned out to be a more incentive-compatible form of rate-of-return regulation rather than a radical departure from it (Lee and Jouravlev, 1997).

In the end, despite, its limitations in practice, price-cap regulation has proved to be less vulnerable to the 'cost-plus' inefficiency and the Averch–Johnson effect (Beesley and Littlechild, 1989). It has also shown that it provides strong incentives to the regulated firm to produce with the cost-minimising input mix, to invest optimally in cost-effective innovation and to innovate in production technology and service offerings, and to adjust optimally to changes in cost. This has happened because the regulated firm retains immediately the benefits of any increase in profits derived from cost savings, even if the regulator later reduces these. It also results, of course, in the companies bearing the costs of inefficient performance and taking a greater part of the financial risk.

The success of the price-cap system of regulation depends on the ability of the regulator to measure the real costs of the companies, choosing the correct periods between reviews as the incentives decline with time because too short a period does not permit companies to innovate (Vickers, 1991).

Restraint on the part of the regulator also allows the companies to enjoy the benefits of their increased efficiency.

The incentive qualities of price-cap regulation also mean that profits and losses can diverge significantly from normal levels. This can be a serious problem because there are many indications that, for political and other reasons, regulated companies will never be allowed to earn excessive profits, even though these profits will lead to future reductions in prices (Helm, 1994a). The experience of British regulators would seem to suggest that public acceptability of price-cap regulation will depend both on transparency and willingness by companies to share benefits with their customers at an early stage.

Whatever regulation mechanism is adopted, success will inevitably depend on regulatory discretion, provided that it is carefully exercised and that the regulators explain the reasons for their decisions. In the United Kingdom, for example, industry-specific regulatory bodies enjoy a great deal of discretion in the performance of their duties and the Office of Water Services has relied on this to limit the accumulation of abnormal profits by the water companies during periods of formal regulatory lag. However, this has led to significant inconsistencies across industries and a widespread feeling of lack of accountability (Ergas, 1994).

Price-cap regulation does effectively control the prices of dominant firms where the market controls their profits. Other factors, however, must be considered, given the relatively high degree of uncertainty that is bound to exist on actual costs and the possibilities for their reduction. The prices set by the regulator must depend in part on actual costs so pure price-cap systems are unlikely to be politically feasible, particularly if regulators are more concerned with consumers' surplus than with the profits of regulated firms (Schmalensee, 1989).

A strong argument for the use of the price-cap system is the lighter regulatory burden. Less regulation leads to both less scope for regulatory capture and a reduction in administrative and compliance costs (Littlechild, 1983). There is a need, however, to take account of 'rate of return considerations in setting and resetting' price-caps (Littlechild, 1988). This considerably increases the information requirements for effective regulation and blurs to some extent the distinction from rate-of-return regulation.

Where the price-cap system has been applied, as in England and Wales and in Chile, where a similar system is used, the public authorities quickly came to the conclusion that in a capital-intensive monopoly industry, such as drinking water supply and sewerage, price controls must be complemented by an assessment of capital expenditure requirements. As a result, British water regulators found themselves 'dragged into a complex mass of detail –

covering the intricacies of business plans and fixing the cost of capital and the value of shareholders' assets' (Helm, 1994a). However, the price-cap system has avoided the lengthy price reviews and the huge legal bills that characterise rate-of-return regulation in the United States.

In particular, there are a number of reasons for preferring price-cap regulation in the periods immediately following privatisation, including, principally, the immediate large potential productivity gains. The gains will be highest in industries where the underlying rate of change in technology and market conditions is faster, as in electricity generation, but there have been spectacular gains even in drinking water supply and sewerage companies. In industries with a decentralised structure, price-cap regulation allows a regulator to generate superior information and overcome asymmetry of information through benchmark or yardstick competition.

Not every aspect of a firm's behaviour can be controlled through the regulation of prices. There are many aspects of potential business conduct in which the holder of a natural monopoly may be able to make economic gains from the power this gives. In consequence, regulators must also regulate these other aspects. To a certain extent, such regulation can be achieved simply by subjecting the companies holding natural monopolies to the basic principles established in competition and antitrust legislation. This requires, of course, a minimum level of competence on the part of the arbitration and judicial systems. In its absence, more direct intervention will be required.

The Problem of Asymmetric Information

One of the common problems facing all regulators is the asymmetric nature of the availability of information to them and to the companies being regulated. The regulator must establish mechanisms to permit, at least, the amelioration of this problem. The normal flow of information required by company law will normally not be sufficient. Fortunately, a number of expedients are available to regulators to aid them in improving their knowledge of the industry they are regulating.

The regulator's information problem is particularly acute in industries where the underlying rate of technological change is high, because knowledge can become obsolete very quickly. It is also a concern in industries where there is only one firm or only a few firms in an industry, where the firms differ substantially from each other (Beesley and Littlechild, 1989). Even where relevant demand and cost structures are observable in principle, the detailed and sophisticated knowledge that a firm has cannot possibly be matched by most regulatory bodies, particularly given the resources and manpower normally available to them. For example, the regulator in Chile has a total staff of 119 and a budget of less than US$ 4 million to supervise an industry with an annual turnover of some US$ 450 million (SSS, 1998).

A regulator can, to an extent, observe the level of costs incurred by companies, as well as the level of their earnings, but it is difficult to monitor any cost-reducing effort, which is the key to productive efficiency and a prerequisite for effective regulation. The results of regulator ignorance are imperfect incentives and impaired economic efficiency. When a regulator is uninformed about industry conditions, then any decision on its part can leave firms with undesirable rents, due to their monopoly of information (Armstrong, Cowan and Vickers, 1994).

The existence of asymmetry in information is a further argument for price-cap regulation schemes rather than the more command-and-control methods of rate-of-return methods. It is likely to prove to be more effective to adopt the goal of designing incentive mechanisms to motivate a firm to employ its superior information to maximise society's objectives while pursuing its own self-interest, rather than to extract rents from its monopoly of information.

Except where regulatory goals are very specific or broad-based performance measures are insufficiently sensitive or excessively variable, it might be advisable to avoid targeting specific components of cost and service quality in incentive schemes. Targeting specific components risks – because the firm has better information about how best to achieve a broad goal relative to the regulator's information – distracting the firm from pursuing those cost-reducing activities for which it does not receive explicitly targeted rewards.

These issues confirm the conclusion that price regulation is likely to be most effective in industries where the extent of the information asymmetry between regulator and firm is small or where the regulator can reduce the firm's informational advantage and acquire adequate information without undue difficulties. Industries that satisfy these conditions are usually those where underlying technological and market conditions change more slowly, as in drinking water supply and sanitation. The slow pace of change means that the regulator can gradually acquire more relevant information to permit it to set realistic efficiency targets. The other industries satisfying the condition are those where it is possible to apply benchmark or yardstick comparisons because there are many firms in an industry and the regulator can use the performance of one as an indication of what another could achieve (Beesley and Littlechild, 1989).

Many of the benefits of private sector participation in water-related public utilities result from the provision of protection for necessary, but politically dispensable, water-related investments from general budgetary pressures. It also provides a means of tapping the greater pool of private capital to help finance them. Although the direct object of regulation is usually pricing policy, the effect of regulation on social welfare, however, depends critically on the investment behaviour that price regulation induces (Vickers and Yarrow, 1988). Given the nature and technological characteristics of most water-related goods and services, the advantages from competition in production and supply are likely to be small unless there is, also, competition in investment.

Where prices are regulated, regulatory agencies must monitor carefully the capital and maintenance spending of the regulated firms to ensure that they make the investments allowed for in the price limits on time and achieve the expansion, quality and other targets for which the investments had been approached. The need for close monitoring is underlined by the capital-intensive nature of most water-related public utilities, which provides scope to evade the constraints imposed by price regulation. This is done through reorganising the firm's investment profile to enhance short-term financial performance at the possible expense of longer-term efficiency and prospects (Bishop and Kay, 1989).

REGULATORY COMMITMENT, RISK AND PRIVATE INVESTMENT

Globally, there is no shortage of capital to make the necessary investments in the water sector. An adequate supply of private finance to the privatised water sector will only be forthcoming if investors are confident that their investment will not disappear. This can occur either though direct expropriation or through an accumulation of regulatory actions that are tantamount to a *de facto* expropriation. Investors also wish to earn a rate of return on the capital invested commensurate with the risk. Potential investors need government commitment to respect, over the long run, their property rights, the rules and regulations governing tariffs, entry conditions, and expansion plans.

The problem of commitment 'arises from a fundamental asymmetry: The regulated price is flexible but the regulated firm's capital stock is not' (Besanko and Spulber, 1992). Although an incentive to act opportunistically exists in any long relationship:

> opportunism may be more characteristic of the policies of public agencies than of private parties because, although courts will prohibit inefficient breach by private

parties, they generally will not proscribe revisions of policies by regulatory or administrative agencies. Instead, courts tend to restrict their review to procedure, process, and consistency (Baron and Besanko, 1987).

Limited commitment powers involve both political and regulatory risks. Political risk arises from potential future radical changes in general government policy and regulatory risk arises from uncertainty surrounding the existing regulatory rules and regulatory environment. A particular feature of both regulatory and political risk is that, unlike other risks, investors perceive them as being almost entirely negative in nature.

Given these circumstances, the potential for under-investment is significant because water-related utilities are highly capital-intensive, and most assets are specialised, tied to a specific location, and extremely durable with slow capital depreciation. Much of the investment is long-term and sunk and this can create a temptation for a regulator to ensure, once capital is irrevocably sunk, that prices are kept artificially low, so that they only cover future avoidable costs, marginal operation and maintenance costs and the return on non-specific assets. This means that there would be no profit margin left to compensate the firm for its prior investment. Under such a situation, a company may be willing to continue operating, but not to invest at efficient levels, because leaving would not allow it to recover any of its investments, while shutting down and deploying its assets elsewhere always involves additional costs. The existence of asymmetry of information between regulator and firm, however, may mean that the regulator is unable *ex post* to set price equal to average avoidable costs and this could mitigate incentives to under-invest.

Such opportunism on the part of regulators can take various forms, but their effect is always to claw back company earnings reducing the value of the initial regulatory contract. Some of the more notable options open to regulators include: interim price reductions; excessively slow depreciation; arbitrary changes in regulatory lag, quality or other aspects of service enhancements, without compensating price increases; disallowing recovery of supposedly 'imprudent' investments; increased investment requirements, without compensating price increases; and market-share reduction (Salant and Woroch, 1992; Helm, 1994b).

This type of confiscatory action by regulators may be profitable for a government where:

1. the direct costs, such as loss of reputation or lack of future investments by utilities, are small compared to the short-term benefits, such as achieving

re-election by reducing utilities' prices or attacking a (foreign-owned) monopoly;

2. the indirect institutional costs, such as disregarding the judiciary, are not too large and if the government's horizon is relatively short (Guasch and Spiller, 1994);

3. the utilities' customers constitute a large proportion of the population, and are mostly captive and outspoken;

4. water-related projects, because of their economic, environmental and social implications and because they serve a lot of people, are highly visible and often serve as powerful political tools which can make a difference in elections (Guasch and Spiller, 1994; Lyon, 1995).

It is essential, therefore, for governments to develop a stable regulatory environment to encourage and maintain private investment in water-related services. Unless there is a stable regulatory environment, the rational fear of *ex post* opportunism by governments will deter efficient investment in sunk cost assets of the type associated with the water sector. The magnitude of the bias to under-invest depends on the nature of the assets involved, the speed of depreciation, the rate of discount and the method used to finance investment, among other factors.

Since the costs of investment are in part determined by the risk involved, any uncertainties associated with regulatory policy will raise the cost of capital to regulated firms. The immediate effect will be an upward pressure on tariffs. Uncertainty will also affect both the magnitude and composition of the investment programme, including the extent of technical innovation. This could also lead to an inefficient technology choice on the part of private investors, providing a bias towards less capital-intensive types of technology. Such uncertainty can also encourage the firm to seek to improve its short-term performance at the expense of the long-term one.

If a government cannot commit itself to a confined regulatory mechanism, private enterprise will not perform better in terms of the public interest than does public enterprise, because risk-averse managers would have an incentive to make as much profit as possible, but not to invest. If governments want to motivate private investment, it is necessary to design institutional arrangements to limit their own ability to behave as opportunist rent-seekers. Otherwise, the public sector might have to assume responsibility for investment. In these cases, service, management and lease contracts become the appropriate, but second best, form of private sector participation.

The factors increasing the possibility of rent-seeking behaviour include setting duties independently of cost considerations, not establishing a clearly defined set of objectives and priorities against which to measure regulatory efficiency, defining regulatory responsibilities in an unduly complex or vague way, or adopting a piecemeal case-by-case methodology which makes every intervention a special case, with regulators given wide powers to adjudicate. It follows that, in designing regulatory regimes, there is a premium on the clarity of objectives for regulators to limit their discretion and the pursuit of their own or interested parties' informal agendas. It is also advisable to rely on general rules, on setting constraints to the scope of regulation, on limiting the total resources available to bureaucracies, and on simplicity where outcomes are closely defined through rules (Helm, 1993). It is also useful to create open and transparent regulatory processes with opportunities for participation by all interested parties.

The only secure route to private sector confidence is a history of rational government committed to policies encouraging private investment in public services. Governments must demonstrate that they do not indulge in *ex post* opportunism. There are, in fact, some policies which governments can adopt to reinforce private sector confidence. For example, deregulating to the maximum, by allocating decision-making authority to the firm responsible for most of the specific investments. This has the advantage of reducing the possibility for government or regulatory failure, but its disadvantages are giving room to any inefficiency arising from market failures. The efficiency of this approach will also depend on the strength of the commitment not to increase regulation in the future.

Probably, the most effective alternative is to start with long-term regulatory contracts which guarantee the right of the regulated firm to earn a 'fair' rate of return on investments. To be effective, these contracts must be credible and backed by guarantees limiting detrimental modifications. This requires setting out the rules of the regulatory scheme in detail and ensuring that past regulatory promises are honoured in future proceedings. Even so, however, it is difficult to ensure regulatory stability. The basic obstacle is that neither regulators nor governments in general can impose binding obligations on their successors.

In Chile and Argentina, for example, very specific regulatory laws have been adopted and regulatory discretion has been strictly limited. The result has been that the private sector has been investing massively in public utilities, particularly in all segments of the electricity sector, but in drinking water supply and sanitation, as well. It is, however, generally difficult to write and enact very specific regulatory laws, in part because all contingencies cannot be effectively anticipated. In addition, having the rules set out in detail limits flexibility in regulation, denying administrative discretion to the regulator. It means increasing costs due to the reduced flexibility to adapt to changing circumstances. It may also encourage political interference in the regulatory work.

Another alternative is to specify in the operating licence or other contract documentation all regulatory procedures and parameters. The use of contract law requires a capable and independent judiciary to arbitrate disputes between the government and the utility. Specifically, courts must view licences as contracts, be willing to uphold them against the wishes of the executive and not grant it too much freedom in their interpretation (Guasch and Spiller, 1994). The use of contract law has important advantages in that it provides a strong guarantee against opportunistic behaviour and it provides a means to institute regulation gradually, adapting the regulatory framework to changing conditions and needs, a factor particularly important in countries with little experience in formal regulation.

The main disadvantage with using contract law is that it, too, may introduce rigidities in the regulatory system. Contracts give the operator substantial bargaining power, limiting the flexibility of the regulatory framework, which would be undesirable if there is a genuine need to amend the licence. Going too far down the road to explicit contract would in effect transfer the responsibility for regulation to judges or tribunals, entrusting the responsibility for determining the return on capital to the judiciary, a task for which it tends to be poorly equipped. Reliance upon project-specific rules embodied in a contract also carries dangers if a government lacks the skills and bargaining leverage to ensure that the resulting contract fairly balances public and private interests.

The use of contract law as a means of restraining regulators and ensuring regulatory credibility is more appropriate when privatisation is limited in scope. Generally, this will be where there are relatively small concessions, build, operate and transfer projects and independent power producers or producers which operate under a contract with state-owned utilities, but it is much less appropriate when privatisation is more comprehensive.

Credibility can also be enhanced through the use of domestic or international guarantees and by building up regulatory reputation and policy credibility. The regulator can overcome the credibility problem by building up a reputation for fair treatment, including a 'fair' rate of return on investments. The longer a government is involved in contracts with private investors, the more the incentives to exploit the sunk nature of their investment will be reduced by the influence of a good regulatory reputation.

External guarantees through the participation of multilateral credit agencies, such as the World Bank and the regional development banks, can help address the problems of limited commitment powers of governments and regulators. Although external guarantees can help reduce regulatory risk, they may also have unintended side effects that undermine successful private sector participation:

1. guarantees might become excessively broad, undermining the efficiencies obtained from investors bearing the risk of a project's failure;
2. since there is no efficient market for regulatory risk, the guarantee might be difficult to price and, if wrongly priced, might send incorrect signals to investors and increase the costs to consumers;
3. excessive guarantees might reduce, rather than enhance, credibility and;
4. since guarantees are not free and government resources are limited, excessive guarantees might delay, rather than speed private sector participation (World Bank, 1995).

Finally, to ensure private sector confidence, it is important to define the regulator's authorities to leave as little discretionary power as possible. Since social objectives are notoriously difficult to define in operational terms, social obligations imposed on the regulator should be kept to a minimum. It is equally important, however, not to go too far and curtail the powers of the regulator excessively, because the protection of the public interest may require changes in the initial regulatory framework. There is a need to find a proper balance between the legitimate interests of the private operator and those of the public.

The long lives of assets in the water industry mean that the rates of return on new investment will be mainly a function of future regulatory decisions rather than of the decisions made at the time of privatisation (Vickers and Yarrow, 1988). In the absence of clear guidance on the long-term conduct of regulatory policy, the uncertainty associated with future public policies can provide a strong incentive for under-investment.

Properly interpreted, rate-of-return regulation – and its inherent promise that utility investors will earn a 'fair' rate of return – can be viewed as 'a form of long-term, incomplete contract with guarantees against capital expropriation' (Yarrow, 1991). It can be seen, therefore, as a means of commitment that addresses the under-investment problem. On this view, the rate-of-return regulation, as implemented in the United States, is attractive because it entails a commitment which has juridical backing and the historical precedent that a fair return on investment will be earned (Rees and Vickers, 1995). The Averch–Johnson effect reinforces these incentive properties.

Price-cap regulation has serious drawbacks in this respect because it fails to provide long-term guarantees as to the decisions made at the regulatory review. Unless clear guidelines binding the decisions taken during regulatory review to ensure a reasonable rate of return are laid down, or emerge from precedent, the cost of capital will increase and there will be an incentive for the firm to under-invest. These guidelines, however, must necessarily embody an explicit feedback from cost reduction to eventual downward tariff adjustment and this would negate some of the superior incentive properties claimed for price-cap regulation. This underlines the need to design the framework of regulation accordingly (Beesley and Littlechild, 1989). In the privatised water industry in England and Wales, this problem is addressed by the requirement that OFWAT must determine the price controls it sets for water companies by reference to an obligation to ensure a reasonable return on capital (Holtram and Kay, 1994).

Pure price-cap regulation may fail to offer the same type of long-term commitment which the rate-of-return regulation, as implemented in the United States, is considered to provide. The decisive influence on commitment, however, is likely to depend more on the structure and behaviour of institutions, both regulatory and political, than on the form of price regulation *per se* (Rees and Vickers, 1995). In both cases there remains the question of whether political institutions are capable of offering secure long-term commitments to regulated firms.

The incentives for opportunistic behaviour tend to be particularly pronounced when realised returns on incentives are unexpectedly high. This implies that price regulation can be designed to make private investments less susceptible to expropriation. For example, since under price-cap regulation profits can diverge significantly from normal levels, rate-of-return regulation can be expected to provide more protection against expropriation. If price-cap regulation were used, it would be preferable to combine it with some way of sharing high profit levels with clients. Finally, efforts should be made to make the link between increased earning and increased effort, diligence and creativity on the part of the regulated firm, as apparent as possible to the customers. If this can be done then they will understand that both they and the firm are better off under a well-designed high-powered price regulation than they would be in its absence (Sappington, 1994).

The Chilean electric power industry affords an interesting example of an effort to address the problem of ensuring an adequate supply of private finance to a capital-intensive industry through a combination of regulatory and market mechanisms. In Chile, the electricity pricing system provides both for regulation of prices and for large consumers to freely negotiate price with any generation company. The regulated prices for bulk power sold to distributors are set at each point of transfer from the national transmission grid to the distributor or node of the high-voltage transmission system. Node prices cannot vary by more than 10 per cent around the average price in the non-regulated contracts. The intention of the regulation is that the private sector will invest in electricity generation to the extent that new projects provide a return on capital compatible with the level of risk (Bitran and Sáez, 1994). If investments in new generation are not made, the future short-run marginal cost will increase, and this will lead to higher node prices, which in turn will give the incentive to expand capacity when demand increases. Under this scheme, most investment choices and decisions are left to the private sector.

Another important source of regulatory risk, which can give rise to under-investment is environmental policy, including any uncertainty about the potential liability for environmental damages arising in the future or from the past actions of the former state-owned enterprise. Other factors such as policy concerning liberalisation and industrial structure, the risks of nationalisation on unfair terms and anti-competitive behaviour of state-owned enterprises are further sources of risk. These risks can be minimised by credible commitment not to change, to the detriment of the regulated industry, the regulatory regime, which guarantees the right of the regulated companies to earn a fair rate of return on their investments.

The United States Safe Drinking Water Act required the Environment Protection Agency to develop maximum levels for 83 new contaminants by the end of 1989 and to develop at least 25 additional primary standards every three years (Phillips, 1993). These new standards raised a number of problems for water utilities, but perhaps the most important of them 'is the fact that few, if any, of the contaminants were taken into account in designing currently installed plants, with the result that both the ultimate treatment processes and final costs of meeting the new standards over the coming years are unknown' (Phillips, 1993).

A closely related problem to open-ended costs is that of the asymmetric treatment of uncertainty by regulators. Regulators tend to apply a stronger standard of scrutiny to extraordinary gains, forcing the firm to pass on these gains to customers, than to extraordinary losses forcing the firm to bear a large part of the cost of bad decisions (Train, 1991). This asymmetry can distort the firm's incentives and induce it to make decisions in a way that ultimately works against welfare maximisation as the firm may be induced to undertake projects that are too conservative.

Privatisation will significantly increase the discount rate applied to investment projects, as the discount factors used by governments are usually low, because unlike private investors, governments can spread the risk over the entire population. Allowance for regulatory risk may also affect the discount rates used in investment appraisal by private investors. There is a tendency for the public sector, facing lower discount rates, to favour long-life, capital-intensive projects, but the funds available are typically rationed and some projects do not materialise while others come to only a slow conclusion (Kay, 1993). The private sector, while it tends to favour shorter-life, lower-capital cost options, ensures that capital is available for any project that meets the rate-of-return criteria.

Privatisation can affect the choice of technology. For example, a higher rate of discount introduces a bias towards less capital-intensive technologies and fuel choices in electricity generation. Thermal power may become the technology of choice rather than hydroelectric generation. Changes in the structure of capital flows with privatisation are not likely to be limited to hydroelectricity generation. Change can be expected wherever there are less capital-intensive technological solutions and where there is competition from less capital-intensive substitutes. For example, water saving alternatives in irrigation rather than new schemes, and with lesser significance, water transport versus rail, road or air, are other areas where privatisation could produce important changes in the structure of investment.

These considerations do not mean to say that governments should interfere with technology choice. However, there could well be some instances when some form of regulatory intervention is warranted. Their concern should rather be to know if the adopted technology 'is the optimal investment choice or if it is the outcome of a bad regulatory system which has lost any credibility to guarantee reasonable rates of return for long run investments' (Laffont, 1994) or some other deficiency of the regulatory framework. If a government decides to use subsidies to encourage the private sector to follow a specific investment path, attention should be paid to the need to ensure that any subsidies are channelled to the most efficient companies. Moreover, subsidies should not be hidden in preferential prices, regulatory concessions or other privileges, rather they should be awarded in such a way that they are explicit and easily accountable to the public.

Regulatory Capture

Regulators have to make their decisions on the basis of the information available to them and the main source of this information is the regulated firm which controls the information it provides and understands that regulatory constraints depend on this information (Helm, 1994b). Assuming that the regulated firm wants to maximise its profits over time and hence seeks to be confronted with the weakest regulatory constraints, there is a strong incentive to try to bias the regulatory outcome in its favour through strategic manipulation of information (Posner, 1974). For example, a far-sighted firm is likely to decide either not to maximise its cost-reduction effort or to switch between high- and low-effort levels, if increases in the effort level today will lead to more stringent regulatory constraints in the future.

A collateral is that the incentive to exercise pressure in order to affect the regulatory outcomes would be weaker under low-powered incentive schemes than under high-powered ones. The latter tend to leave high potential rents to the regulated industry and thus create high returns to collusion and capture (Laffont and Tirole, 1991). This inference would be true for incentive schemes where the regulatory agency plays an active role in regulation. If the regulatory agency has little discretion, if it restricts its role to that of an accounting office, then stakes in collusion may be reduced by the use of high-powered incentive schemes (Laffont and Tirole, 1990, 1991).

In a regulated industry, the major reasons for regulatory capture are informational asymmetries between the regulated firm and its regulators, between regulators and political authorities and between political authorities and voters. The basic informational asymmetry between firms and regulators gives rise to imperfect incentives, allows firms to extract advantages from the monopoly of information that they enjoy and, therefore, impairs economic efficiency. Regulatory capture does not commonly include direct monetary payments or bribes. The more common forms of influence include the fact that the regulated industry is an important source of future employment opportunities for the regulatory agencies' staff. Personal relationships provide incentives for government officials to treat their partners from the regulated firms generously. The regulated industry may cater to the regulatory agency's bureaucratic desire for a quiet life, or for larger resources, by refraining from publicly criticising agency management. It may also make indirect transfers, such as monetary contributions to political campaigns, as well as the votes and lobbying of employees, shareholders, suppliers or the members of local government agencies.

Generally, an industry will have better access to technical talent than the regulatory agency. This, coupled with the tendency for the better-trained technical staff from the public sector to migrate to the regulated firms, can create a shortage of qualified manpower for the regulator. The specialised nature of the duties exercised by regulators and of the skills they develop in the performance of their duties makes them attractive employees for the regulated industry. Their obvious career path is to move to the regulated private sector where they can claim lucrative appointments. The nature of their business and social contacts reinforces this tendency. This phenomenon creates conflict of interest and distorts their incentives while working for the regulatory agency. It also undermines the independence of the regulating authorities and has a general debilitating effect on the effectiveness of regulation.

Every regulatory agency is required to be closely involved with the regulated companies on a day-to-day basis. This can result in the agency identifying itself or the interests of the public with the aims of the industry. Such collusion between the regulator and the industry can frustrate the objective of controlling and diminishing marker failures. The regulator can become, instead, an advocate of the industry or even an instrument for the maintenance and reinforcement of monopoly power. Weakness of regulatory policy towards effective competition, promotion of new entry, contestability and price and quality regulation reveals such regulatory capture. It leads to higher prices, lower quality and the protection of incumbents from competitive entry. Regulatory capture can also be revealed on the stock market: abnormally high returns associated with changes in the regulatory environment could indicate capture by the regulated industry, while abnormally low returns could indicate capture by consumer interests.

To avoid these problems, regulatory personnel need not only be technically qualified, but also well paid relative to the salaries in the regulated industry. This had led, in some countries, to a decision to place the regulator's staff outside the restraints of public sector salary scales. In addition, if possible, the staff of regulating agencies should be prohibited from working in the regulated industry for a specified period after their appointments are terminated. However, post-employment restrictions may be costly or even impossible to enforce.

In an ideal world, to be effective and to avoid the problem of regulatory capture, regulatory agencies would have the following characteristics:

1. they would be staffed with people of unblemished reputation and with adequate technical skills capable of fulfilling their functions at the level of expertise and efficiency required to confront the private operator on, at least, an equal basis;
2. they would enjoy consistent political support and, while subject to periodic evaluation, receive a minimum of day-to-day political interference;
3. they would have adequate financial, human and informational resources and an independent budget;
4. they would only possess limited discretion in the discharge of their duties to minimise the risk of regulatory capture;
5. they would open the regulatory process to public scrutiny and would explain and justify at least some decisions and publish the evidence on which they are based;

6. they would be invested with sufficient autonomy to limit the possibility of being captured by particular interest groups, including those within government.

It cannot be expected that regulatory systems anywhere will exhibit all these characteristics. It is, however, of fundamental importance in establishing a regulatory system to limit the scope for regulatory capture. There are always incentives for the regulator to act in the interests of incumbents in the industry rather than in those of consumers or potential rivals. Consequently, there are grounds to argue that it could be desirable to introduce countervailing incentives for regulators to act in the interests of society as a whole. Any reward, however, will probably have to be limited to official recognition, promotion, status and other similar incentives which are used to reward the performance of public servants in general. Other incentives – such as regulatory competition or linking pay to some aspects of company performance – are difficult to implement and are likely to have perverse consequences for the regulated industry and consumers. On the whole, 'it is difficult to see an alternative superior to allowing a disinterested regulator to make a decision – provided that all reasonable precautions have been taken to ensure that the regulator is indeed disinterested' (Holtram and Kay, 1994).

REFERENCES

Armstrong, Mark, Simon Cowan and John Vickers (1994), *Regulatory Reform: Economic Analysis and British Experience*, The Massachusetts Institute of Technology, MIT Press Series on the Regulation of Economic Activity-20, Cambridge, Massachusetts, London, England: The MIT Press.

Averch, Harvey and Leland L. Johnson (1962), 'Behavior of the firm under regulatory constraint', *The American Economic Review*, **52** (5), 1052–1069.

Bacon, Robert (1994), *Restructuring the power sector: the case of small systems*, FPD Note series, Washington D.C: The World Bank.

Baron, David P. and David Besanko (1987), 'Commitment and fairness in a dynamic regulatory relationship', *Review of Economic Studies*, **54** (3), 413–436.

Barraqué, Bernard J. (1993), 'Water management in Europe: beyond the privatization debate', *Economia delle i di Energia e dell'Ambiente*, **34** (3), 43–80.

Baumol, William J., John C. Panzar and Robert D. Willig, with contributions by Elizabeth E. Bailey, Dietrich Fischer and Herman C. Quirmbach (1982), *Contestable Markets and the Theory of Industry Structure*, New York: Harcourt Brace Jovanovich.

Beesley, M.E. and S.C. Littlechild (1989), 'The regulation of privatized monopolies in the United Kingdom', *The RAND Journal of Economics*, **20** (3), 454–472.

Besanko, David and Daniel F. Spulber (1992), 'Sequential-equilibrium investment by regulated firms', *The RAND Journal of Economics*, **23** (2), 153–170.

Bishop, Matthew R. and John A. Kay (1989), 'Privatization in the United Kingdom: lessons from experience', *World Development*, **17** (5), 640–654.

Bishop, Matthew, John Kay and Colin Mayer (1995), 'Introduction', in Matthew Bishop, John Kay and Colin Mayer (eds), *The Regulatory Challenge*, Oxford: Oxford University Press, pp. 1–17.

Bitran, Eduardo and Raúl E. Sáez (1994), 'Privatization and regulation in Chile', in Barry P. Bosworth, Rudiger Dornbusch and Raúl Labán (eds), *The Chilean Economy. Policy Lessons and Challenges*, Washington, D.C.: The Brookings Institution, pp. 329–377.

Boadway, Robin W. and David E. Wildasin (1984), *Public Sector Economics*. 2nd. edn, Boston, Toronto: Little, Brown and Company.

Breautigam, Ronald R. and John C. Panzar (1989), 'Diversification incentives under "price-based" and "cost-based" regulation', *The RAND Journal of Economics*, **20** (3), 373–391.

Byrnes, Patricia, Shawna Grosskopf and Kathy Hayes (1986), 'Efficiency and ownership: further evidence', *The Review of Economics and Statistics*, **68** (2), 337–341.

Crew, M.A. and P.R. Kleindorfer (1985), 'Governance costs of rate-of-return regulation', *Journal of Institutional and Theoretical Economics*, **14** (1), 104–123.

Dnes, Antony (1995), 'Post-privatization performance – regulating telecommunications in the U.K.', *Viewpoint*, Note 60, The World Bank, Private Sector Development Department, Vice Presidency for Finance and Private Sector Development.

Ergas, Henry (1994), 'Comment on "Appropriate regulatory technology", by Jones', in *Proceedings of the World Bank Annual Conference on Development Economics*, 1993, Washington, D.C.: The World Bank, 205–207.

Feigenbaum, Susan and Ronald Teeples (1983), 'Public versus private water delivery: a hedonic cost approach', *The Review of Economics and Statistics*, **55** (4), 672–678.

Freeman, Andrew (1991), 'Fewer than expected', *Financial Times*, 22 November.

Guasch, J. Luis and Pablo T. Spiller (1994), *Regulation and Private Sector Development in Latin America*, Washington D.C.: The World Bank.

Guislain, Pierre (1992), *Divestiture of state enterprises. An overview of the legal framework*, World Bank Technical Paper No. 186, Washington, D.C.: The World Bank.

Helm, Dieter (1993), 'The assessment: reforming environmental regulation in the UK', *Oxford Review of Economic Policy*, **9** (4), 1–13.

Helm, Dieter (1994a), 'Price limits do not hold water', *The Times*, 29 July.

Helm, Dieter (1994b), 'British utility regulation: theory, practice, and reform', *Oxford Review of Economic Policy*, **10** (3), 17–39.

Holtram, Gerald and John Kay (1994), 'The assessment: institutions of policy', *Oxford Review of Economic Policy*, **10** (3), 1–16.

IEA (International Energy Agency) (1994), *Electricity supply industry. Structure, ownership and regulation in OECD countries*, Paris: Organisation for Economic Co-operation and Development.

Izaguirre, Ada Karina (1998), *Private Participation in the Electricity Sector – Recent Trends*, Public Policy for the Private Sector, Note No. 154, Washington, D.C.: The World Bank.

Jones, Leroy P. (1994), 'Appropriate regulatory technology. The interplay of economic and institutional conditions', in *Proceedings of the World Bank Annual Conference on Development Economics*, 1993, Washington, D.C.: The World Bank, pp. 181–194.

Kay, John (1993), 'Efficiency and private capital in the provision of infrastructure', in *Infrastructure Policies for the 1990s*, Paris: Organization for Economic Co-operation and Development, 55–73.

Kessides, Christine (1993), *Institutional options for the provision of infrastructure*, World Bank Discussion Papers, 212, Washington, D.C.: The World Bank, .

Laffont, Jean-Jacques (1994), 'The new economics of regulation ten years after', *Econometrica*, **62** (3), 507–537.

Laffont, Jean-Jacques and Jean Tirole (1990), Accounting and collusion, mimeo, M.I.T., as quoted in Laf and Tirole (1991).

Laffont, Jean-Jacques and Jean Tirole (1991), 'The politics of government decision-making: a theory of regulatory capture', *The Quarterly Journal of Economics*, **106** (4), 1089–1127.

Lambert, David K., Dimo Dichev and Kambiz Raffiee (1993), 'Ownership and sources of inefficiency in the provision of water services', *Water Resources Research*, **29** (6), 1573–1578.

Lee, Terence and Andrei Jouravlev (1997), *Regulation of the Private Provision of Public Water-Related Services*, LC/R.1635/Rev. 1, Santiago, Chile: United Nations, Economic Commission for Latin America and the Caribbean.

Lewis, Tracy R. and David E.M. Sappington (1988), 'Regulating a monopolist with unknown demand and cost functions', *The RAND Journal of Economics*, **19** (3), 438–457.

Littlechild, S.C. (1983), *Regulation of British Telecommunications Profitability*, London: HMSO.

Littlechild, S.C. (1988), 'Economic regulation of privatised water authorities and some further reflections', *Oxford Review of Economic Policy*, **4** (2), 40–67.

Lyon, Walter A. (1995), 'Privatization law and water institutions', in American Society of Civil Engineers/Economic Commission for Latin America and the Caribbean, (1997), *Proceedings of the Workshop on Issues on the Privatization of Water Utilities in the Americas*, Santiago, Chile, 4–6 October 1995, LC/R. 1722/Add.1, Santiago, Chile: United Nations, Economic Commission for Latin America and the Caribbean, 45–51.

Murray, Alasdair (1995), 'OFWAT demands separate quotes', *The Times*, 14 December.

Nakamoto, Michiyo (1991), 'Corporate differences highlighted', *Financial Times*, 22 November.

OFWAT (Office of Water Services) (1995a) *Water regulator sets out his information requirements*, PN11/95, Birmingham, England: Office of Water Services.

OFWAT (1995b), *OFFER and OFWAT issue joint consultation paper on proposed acquisition of SWALEC by Welsh Water plc*, PN 38/95, Birmingham, England: Office of Water Services.

OFWAT (1995c), *Water regulator calls for separate listing for merged businesses*, PN 39/95, Birmingham, England: Office of Water Services.

Paredes, Ricardo D. (1995), 'Evaluating the cost of bad regulation in newly privatized sectors: the Chilean case', *Revista de Análisis Económico*, **10** (2), 89–112.

Perry, Martin K. (1984), 'Scale economies, imperfect competition, and public policy', *The Journal of Industrial Economics*, **32** (3), 313–384.

Perry, Martin K. (1989), 'Vertical integration: determinants and effects', in Richard Schmalensee and Robert D. Willig (eds), *Handbook of Industrial*

Organization. Vol. I, Amsterdam: North-Holland, Elsevier Science Publishers, pp. 183–255.

Phillips, Charles F. (1993), *The Regulation of Public Utilities. Theory and Practice*, Arlington, Virginia: Public Utilities Reports.

Posner, Richard A. (1974), 'Theories of economic regulation', *The Bell Journal of Economics and Management Science*, **5** (2), 335–358.

Rees, Ray and John Vickers (1995), 'RPI-X price-cap regulation', in Matthew Bishop, John Kay and Colin Mayer (eds), *The Regulatory Challenge*, Oxford: Oxford University Press, pp. 358–385.

Salant, David J. and Glenn A. Woroch (1992), 'Trigger price regulation', *The RAND Journal of Economics*, **23** (1), 29–51.

Sappington, David E.M. (1994), 'Designing incentive regulation', *Review of Industrial Organization*, **9** (3), 245–262.

Schmalensee, Richard (1974), 'Estimating the costs and benefits of utility regulation', *The Quarterly Review of Economics and Business*, **14** (2), 51–64.

Schmalensee, Richard (1989), 'Good regulatory regimes', *The RAND Journal of Economics*, **20** (3), 417–436.

Schmalensee, Richard (1995), 'What have we learned about privatization and regulatory reform?', Revista de Análisis Económico, **10** (2), 21–39.

SSS (Superintendencia de Servicios Sanitarios) (1998), *Informe de Gestión del Sector Sanitario, 1996–1997*, Santiago, Chile: Superintendencia de Servicios Sanitarios.

Stigler, George J. (1971), 'The theory of economic regulation', *The Bell Journal of Economics and Management Science*, **2** (1), 3–21.

Torres, Clemencia (1995), 'Regulatory schemes and investment behavior in transmission of electricity: the case of Argentina', *Revista de Análisis Económico*, **10** (2), 203–235.

Train, Kenneth E. (1991), Optimal Regulation. *The Economic Theory of Natural Monopoly*, Cambridge, Massachusetts, London: The MIT Press.

Vickers, John (1991), 'Government regulatory policy', *Oxford Review of Economic Policy*, **7** (3), 13–30.

Vickers, John (1995), 'Concepts of competition', *Oxford Economic Papers*, **47** (1), 1–23.

Vickers, John and George Yarrow (1988), *Privatization: an Economic Analysis, The Massachusetts Institute of Technology*, MIT Press Series on the Regulation of Economic Activity 18, Cambridge, Massachusetts, London: The MIT Press.

Vickers, John and George Yarrow (1991), 'Economic perspectives on privatization', *The Journal of Economic Perspectives*, **5** (2), 111–132.

World Bank (1992), *The Bank's role in the electric power sector. Policies for effective institutional, regulatory and financial reform,* Washington, D.C.: World Bank.

World Bank (1994), *World Development Report 1994,* New York: Oxford University Press.

World Bank (1995), *Bureaucrats in business. The economics and politics of government ownership,* A World Bank Policy Research Report, New York: Oxford University Press.

Yarrow, George (1991), 'Vertical supply arrangements: issues and applications in the energy industries', *Oxford Review of Economic Policy,* **7** (2), 35–53.

6. Water Management and the Challenges of the 21st Century

This book has taken an approach towards water management that defines its central tasks as one, ensuring the efficient allocation of water and two, ensuring that, once allocated, water is efficiently used. This may not be a revolutionary, or even new proposal, but it is very different from the traditional approach which sees the prime objective of water management as the control of water through the construction of civil engineering works. Moreover, it is also argued that the secondary objective of water management should be the establishment of a policy and regulatory framework to allow the private sector to take responsibility for most water management actions.

THE ALLOCATION IMPERATIVE

It may be claimed that to see the allocation of water as the central purpose of water management is a very particular view, an economist's interpretation of its nature. It seems to the author, however, that, unless water is efficiently allocated and efficiently used, then we have no criteria for judging whether it is in scarcity or surplus and, therefore, of knowing how or whether its use should be controlled. It is exactly this failure to understand what the objective of water management should be that in the past led to an over-commitment to attempts to the use the construction of water control works as a foundation for economic and social development. There is a common perception in the literature that '(R) river basins tend to be heartlands of critical importance for the success of development programmes' (Scudder, 1994). Such exaggeration of the potential role of civil engineering works neither helps to manage water rationally nor leads to an understanding of the real role that water management can play in tackling economic and social issues.

Along with the emphasis on works and development, for many years there has been, within the water management profession, a concern with the efficiency of management activities. This has been particularly true on the part of financing agencies both within governments and by the multilateral development banks. This concern has not, however, been focused on the question of the allocation of water among uses, the macro issue, but rather on the micro question of how to build water control projects more efficiently,

In the literature of project analysis and evaluation there are two widely accepted principles mentioned in the quotation below. These are, first, to correctly price water services so as to ensure cost recovery and, second, to ensure the internalisation of external costs caused by pollution and other side-effects of water projects. However, the application of these principles in project analysis has not been able to resolve the issue of the misallocation and inefficient use of water because there has been no link between project evaluation and project implementation.

> The appraisal approaches used at the Bank and at other funding agencies are based upon the generally accepted principles outlined voluminously in the literature ... The application of the principles is flawed, because when faced with economic scarcity, the water itself must be priced at its opportunity cost. This is typically not done and leads to serious misallocation of water between different uses of water (Rogers, 1992).

What is missing are the connections provided by the market and which are explicit in decisions on investments and services by the private sector. Moreover, there are serious, if not insurmountable, difficulties in allocating water among sectors efficiently in the absence of a market. This is why in this book, I have argued for using the market for water allocation and for introducing market principles into the provision of water services through transferring the responsibility for them to the private sector.

Only once this is done can any rational judgement be made on the issue of the state of the supply of water. Today we are faced not with water scarcity but with confusion and uncertainty caused by the misallocation and misuse of water. It is this that must be put right first, before any other decisions can be taken. This book is intended to be a small contribution to the effort required to rectify this existing confusion and uncertainty about what issues water management should be addressing.

Getting the Price Right

Much of the water management literature has put considerable emphasis on the need to get the price of water right, both for water as a resource and for the services derived from its use. Reference has been made to this literature in the earlier chapters. There seems to be an inhibition, however, when the logic of these arguments is taken to its conclusion, resulting in a clear paradox that is far from being resolved. The discussion tends to be confused on the nature of water as an economic good, even when it is proposed that water should be priced as other commodities are priced. It is necessary, therefore, to be quite clear that water as a natural resource is not a 'public good'. It is an economic commodity. It is only a public good when it forms part of a landscape, which in turn has been declared to be a public good. It has this characteristic in common with land and should be treated accordingly. If water did not flow and had not been originally so abundant in nature then, perhaps, we would not now remain so confused about its nature as an economic good.

It is also often claimed that water is different from other resources because of its distinct character in that it naturally flows. In markets, however, all goods tend to flow and there is no problem in isolating specific quantities in time and space. This is just what is done when water is used to drink, to irrigate or to generate energy. From my own personal experience as an owner of a very small water-right in Chile, I know that it is possible to use for my own purposes at the given time the flow of water in the canal that passes through my land. At that moment the water is mine and, within the law, I may use it as I see fit. Once the time allocated to me ends then the water flows on to be used by one of my neighbours. There is no particular difficulty in managing this system that treats water as private property because there is a properly constituted institutional framework in which to do so.

The issue of the nature of water as an economic good lies not with the nature of water, but with the inheritance of the belief that water is a free good. Until we can overcome this 'terrible' prejudice, water management will not be able to dedicate itself to ensuring that the water resource can maximise its contribution to economic and social development as we enter the 21st century.

GIVING THE PRIVATE SECTOR ITS HEAD

The provision of water-related services has been marked in many parts of the world by massive government failure. Government failure has not been universal, there are examples of the successful public sector administration of services, particularly in electricity generation, but they are the exception. The occurrence of serious and pervasive government failure is, without doubt, the fundamental reason for the current widespread interest in the privatisation of the provision of water-related services.

There is now a movement underway in nearly all countries to increase private sector participation in water management through transferring water-related services into private hands. In irrigation this tendency is all but universal, as former publicly managed irrigation districts are transferred to associations of farmers. This privatisation process usually does not require any regulation, as irrigation farming is everywhere subject to market competition. In electricity generation, the sale of generating facilities is widespread. In this case too, extensive regulation can be avoided if, in structuring the industry, competition can be introduced. It is the involvement of the private sector in drinking water supply and sanitation, due to the limitations placed on competition by the technological nature of the industry, which will definitely require the development of effective regulatory structures.

Regulation

In regulatory policy, there are two clear priorities, the development of an effective regulatory capacity and the establishment of the independence of regulatory authorities, free from direct political interference. At the same time, it is necessary for those defining regulatory policy to be absolutely clear as to what its real objectives must be. The objective of regulation is to ensure that in activities subject to natural monopolies there is the introduction of the maximum competitive pressure. Where competition is not feasible, regulators must act as a substitute for the market, taking on some of the functions of competitors, attempting to provide similar incentives to improve efficiency by regulating aspects of the firm's conduct. It is not the objective of regulation to manage the companies that own, or operate under franchising arrangements, water-related public utilities. It is for this reason that the functions of the regulator must be clearly defined and strictly limited to the absolutely necessary. In general, that is why price regulation is the preferred basic approach, leaving all other decisions to the utility managers.

In defining the type of price regulation, notwithstanding the criticisms that have been made of its incentive properties, rate-of-return regulation does offer some advantages. It can provide potential investors with a solid guarantee of a fair rate of return, it offers a type of long-run commitment to returns which is crucial for investments with a high sunk-cost component, as in hydroelectricity generation and drinking water supply and sewerage. It provides weaker incentives for cost reduction, but performs well under uncertainty, important in economies with histories of high inflation and general macroeconomic instability, and should have a downward impact on the cost of capital. At the same time, it reduces the ability of the regulated firm to profit from regulatory ignorance or favourable cost shocks, important in countries with little regulatory experience.

There are, however, stronger reasons for preferring price-cap regulation, particularly immediately following privatisation. Productivity gains, which price-cap regulation encourages, are potentially largest at this moment. Gains are also potentially larger in cases where changes in the technology applied and in market conditions will be faster, but most state-run public utilities are characterised by low productivity. In electricity generation, for example, it is common that the prior existence of under-investment and poor management has created a technological gap. Price-cap regulation is also attractive in settings in which, through technological change, competition is increasing. In the end, however, it seems likely that the overall incentive effects of price-cap and rate-of-return approaches do not differ much because both forms of regulation, at least as typically implemented, have some features of the other.

Under the right conditions, commercial code regulation may be a useful complement to, or substitute for, other forms of regulation, particularly in smaller countries and in those industries where effective competition is feasible. This form of regulation is simple to implement, is very inexpensive, and provides a means to institute regulation gradually, all factors particularly important in countries with little experience in formal regulation. It would also be useful to strengthen the use of the commercial code or anti-monopoly legislation by encouraging consumer participation in the oversight process and by encouraging the organisation of small groups of customers into larger, more effective bargaining units.

Adequate information is of paramount importance for effective regulation. The regulated company's management always knows far more than the regulatory agency about both industry costs and demand conditions. Asymmetric information allows a firm to extract rents from its monopoly of information resulting in an overall welfare loss. The existence of the informational asymmetry suggests that:

1. the regulatory goal should be to design incentive mechanisms for the regulated firm that will induce it to maximise society's objectives while pursuing its own self-interest;
2. the prospects of generating information for regulatory purposes should be an important consideration in a government's decision on the nature of the regulatory regime and the structure of an industry.

The most promising path to take in order to effectively address the problem of asymmetric information in privatised water-related utilities seems to be some form of benchmark competition. The advantages of benchmark competition are part of the case for having a horizontally separated rather than national structure in water-related public utility industries.

Given that competition is at present limited in the core transportation and distribution services of the water sector, that markets for its services are characterised by informational asymmetries between service providers and their customers, and that service providers usually operate under regimes of monopoly, it is extremely unlikely that price control alone would be capable of giving sufficient incentives to profit-maximising firms to make socially optimum quality choices. Unfortunately, the regulation of service and product quality is one of the most neglected problems in the debate on private sector participation in the water sector. There is, therefore, a strong case for supplementing price regulation with the regulation of service quality. The most promising approach to this problem would seem to be customer compensation schemes, guaranteed standards of performance and minimum quality of service standards. Formal incorporation of quality of service measures in price regulation is another interesting approach, but it can be difficult to implement in practice.

Where a public utility faces separate regulators for quality of service, pollution and other environmental aspects, and prices, whose preferences for the various possible actions typically conflict, what is generally referred to as the common agency problem can arise. This can create tension between regulators and the danger of inefficient outcomes as well as of inconsistency and a lack of credibility. These considerations underline the need for close co-operation between the regulatory agencies and decision-making authorities, and for an explicit duty to be imposed on them to balance the costs against the benefits of regulatory decisions.

Private Sector Participation

At present, most countries rely on administrative or 'command and control' means for water resource and environmental management. The transfer of responsibilities from the state to the private sector will produce a need – this need is already detectable in those countries which have advanced most on the road of privatisation – for greater reliance on prices and other incentives to encourage efficient use and allocation of water. It will also require greater user participation in water resource management.

Many of the benefits of private sector participation in water-related pubic utilities result from the provision of protection to necessary, but politically dispensable, water-related investments from general budgetary pressures. It also provides a means of tapping the greater pool of private capital to help finance them. This implies that the effect of regulation on social welfare depends critically on the investment behaviour that it induces in regulated firms. An adequate supply of private finance to the privatised water sector will only be forthcoming if investors are confident that their investment will not disappear though direct expropriation or through creeping regulation. Investors also expect to earn a rate of return on the capital invested in the sector that is commensurate with the risk they take. Potential investors need government commitment to respect, over the long run, their property rights, the rules and regulations governing tariffs, entry conditions, and expansion plans. It is essential, therefore, to develop a stable regulatory environment to encourage and maintain private investment in water-related services. Unless there is a stable regulatory environment, the rational fear of *ex post* opportunism by governments will deter efficient investment in sunk cost assets. The only secure route to private sector confidence is a history of rational government committed to policies encouraging private investment in public services. Governments must demonstrate that they do not indulge in such *ex post* opportunism.

One effect of privatisation will be to increase significantly the discount rate applied to investment projects, as the discount factors used by governments are usually low. This means that privatisation can affect the choice of technology. For example, a higher rate of discount implies a bias toward less capital-intensive technologies and fuel choices in electricity generation. Thermal power may become the technology of choice rather than hydroelectric generation. Change can be expected wherever there are less capital-intensive technological solutions and where there is competition from less capital-intensive substitutes. Irrigation *vis-à-vis* rain-fed agriculture, and to a lesser significance water transport, versus rail, road or air, are other areas where privatisation could produce important changes in the structure of investment. If a government decides to use subsidies to encourage the private sector to follow a specific investment path, attention should be paid to the need to ensure that any subsidies are channelled to the most efficient companies and that they do not unduly interfere with the play of market forces.

Organising effective user and consumer involvement in the regulatory process will probably take time, because in many countries there is little experience of such participation. Water management has been characteristically highly centralised within the public sector and in central governments. One consequence has been highly centralised systems for service delivery which have a history of being unresponsive to customer demands and have been subject neither to market nor political tests of responsiveness. As a result, most of the population has never had to face the realities of the fact that water-based services are not free but have to be paid for. Moreover, it is not widely understood that the choice of service level should be made collectively and rationally in light of the costs and benefits to the community at large. Because decentralisation shows that there is such an obvious need for consumer involvement, greater private-sector participation should aid to stimulate greater consumer involvement.

Community participation is often an essential feature of the provision of drinking water supply and sanitation services in rural areas. Public authorities should provide an appropriate legal, institutional and policy framework to promote such participation. Studies clearly show that projects with high participation in project selection and design are much more likely to maintain the systems in good condition than those characterised by more centralised decision-making. Rural drinking water supply programmes should be demand-driven, any subsidies should not distort the community's choice, and beneficiaries should mobilise a considerable portion of the resources. Demand-side assistance, that is giving subsidies directly to households and not to utilities, should be encouraged because these ensure that the intended beneficiaries are properly targeted.

Structural reforms should seek to isolate the natural monopoly elements in an industry and to prevent the firms entrusted with activities with natural monopoly characteristics from extending their monopoly powers beyond the segment of the market where these characteristics exist. Failures to isolate the natural monopoly elements and to create adequate competition can considerably complicate conduct regulation and make its task more demanding and its scope broader than necessary. Any failure will also impose on regulators the task of trying to compensate for the deficiencies in structure through more intrusive conduct regulation. Inappropriate industrial structure is one of the main causes of regulatory failure and competition is the best form of control.

In most countries, centralisation meant water-related utilities were usually heavily vertically integrated to a degree that they included all operational and support functions, including those that do not exhibit natural monopoly characteristics. It is clear that many utilities could realise substantial cost savings and enhance efficiency through both horizontal and vertical separation by means of franchise arrangements with private firms. Many activities can and should be opened to direct competition. The examples are many; in fact almost everything except the overall co-ordination of activities could be contracted out in any system, although in practice such extreme outsourcing is not often practised.

Horizontal separation into geographically discrete companies is almost always one of the accompanying features of any restructuring of any water-related service, whether electricity generation, irrigation or drinking water supply and sanitation. For example, none of the large national public utilities remain untouched by decentralisation in Latin America. The breaking up of large public entities into smaller public or private concerns can be considered a necessary requirement for improving efficiency and introducing competitive pressures, if not actual market competition.

Functional separation is, however, by no means a panacea. Attempts to separate closely interdependent activities can impose high costs on the sector, including the loss of the economies of scale and scope. There are also the costs of sector restructuring and the possible loss of some internalisation of externalities, which need to be carefully weighted against the potential benefits of cost-minimising behaviour under competitive pressure. If these factors are significant, there may be a case for the continuation of vertically integrated monopoly.

THE CONTRIBUTION OF WATER MANAGEMENT TO RESOLVING HUMAN PROBLEMS IN THE 21st CENTURY

The contribution of water management to resolving the major social issues facing human society at the dawn of the 21st century are, strangely enough, or maybe not, very much the same as those that were being faced as the 20th century dawned. The greatest challenge remains, to accomplish what has not been possible in this century, to ensure that water management contributes to the achievement of a universally high level of economic development that is both sustainable and equitable. Within this greater challenge there are three main issues which must be met. These are, in my opinion in order of importance, to increase productivity, to eliminate poverty and to minimise the impact of economic activities on the environment. So the question we must address is what is expected from water management so that people and governments can go about the task of meeting these challenges.

The Role of Water Management in Improving Productivity

Despite optimism at various times and in many quarters, increasing economic productivity is still the major challenge facing the countries of the world. What is true about the economy as a whole holds equally true for those goods and services closely related with the water resource. Most water-based services, not only drinking water supply, but also hydroelectricity generation and irrigation are still run at a loss in most countries and require considerable subsidies not only for capital expansion, but even for routine operations. These subsidies are provided at the expense of alternative, more socially profitable use of these financial resources. If water-based services are to make a maximum contribution to economic growth in general, then they must become financially self-sufficient, including the financing of future capital investment.

Profitability and self-financing without subsidy is, as has been discussed, quite achievable for all water-based services. There are various alternative means that are already being used in many countries to ensure that these services maximise their contribution to economic productivity, to increasing the productive base and, through these means, to economic growth.

Fundamental to this process is the reconsideration of the role of the government and the public sector in the economy and a revival of the faith in markets. Where improvements have been achieved, the most common change has been the transfer through sale or concession of many water-based productive activities to the private sector. This has been particularly the case with hydroelectric power generation and irrigation, but it is growing among water supply and sanitation companies. An even more important institutional change is the decision in some countries to make water rights real property that can be freely transferred without reference to the bureaucracy.

If these improvements are to become commonplace and widespread, then the following steps are necessary:

1. Governments must get out of the provision of basic water services. As has been discussed, many different procedures are available and have been used, but, to summarise, among the most common are:

(a) to transfer responsibility from a central government ministry to another public institution, such as an autonomous public corporation, or to the states or provinces in the federal countries, or to a regional authority or a municipality in countries with a unitary system of government;

(b) to transfer management responsibilities to formally constituted water-user associations, particularly suitable for irrigation and rural drinking water supply;

(c) to grant water services in concession to private companies. This is particularly common for drinking water supply and sanitation, although it is also being considered for irrigation works in some countries;

(d) to privatise through the sales of shares or by tender, a normal practice for hydroelectricity generation, and it is being increasingly applied to water supply and sanitation services.

2. Create water markets. In some countries, water markets have already been established or are being considered through the assignment of property rights to water rights and the permitting of the holders of the rights to freely trade them. Such markets have existed successfully in Chile since the early 1980s.

3. Price water services to cover the total costs of their provision so that saleable water services, for example, drinking water, irrigation, and hydroelectricity, finance the total costs of provision from tariff revenues, including the control of the external or environmental costs associated with their provision.

The continuation, the deepening and widening of these policies is the only basis by which the use of the water resource can become more productive and its role as an economic good can be fully played.

Water Management and the Satisfaction of Basic Needs

The ample evidence of failure to provide adequate drinking water supply and sanitation has justified much of the criticism made about the poor management of water services. In general, there are far too many examples of badly run water services which fail to meet the demands placed upon them to be able to claim that the administration of basic water services does not demand radical reform.

Since the 1960s, tremendous efforts have been made to improve public utilities. Nevertheless, the efforts have consistently failed to achieve the objectives set. One of the principle limitations has been the weak financial situation of the state-owned public utility companies. The lack of financial resources has been complicated in many cases by poor management. These two factors have had the consequence of insufficient increases – and even decrease – in the provision of services and have constituted an important limitation even for those systems that have shown a better performance. There are, in consequence, many arguments for searching for means of improving the management of water-related public services.

No one wishes to pay more for any good or service, but only moving towards self-financing will allow the whole of the urban and rural population to have access to the basic services related to the water resource. Moving towards the self-financing of public utilities is an end-of-century imperative. There should be no doubt in policy-makers' minds that the financial restrictions can be eliminated through the establishment of tariff systems which allow the whole cost of the provision of basic water services to be met, even in the poorest countries. There should be no doubt either, that this is the only way of having well-managed services that are the means to maximise the contribution of water management to the elimination of poverty.

The achievement of self-financing is the basis for one of the strongest arguments in favour of the privatisation of public utilities. The objective is not, however, privatisation for its own sake, but the transformation of public utilities into companies where the investments and the provision of services do not continue to be in deficit and the quality of the services low, especially for the poorest members of the population. Achieving effective and efficient services is perhaps the most urgent priority for water management policy at the end of this century and the beginning of the next.

Water Management and the Environment

In this book little explicit attention has been given to the environmental aspects of water management. However, its importance is not to be ignored. We build dams, drain swamps, make diversions and otherwise use water for the production of electricity, irrigation, flood protection, drinking water supply and only incidentally are we aware that we have changed the environment. Yet we all know that freshwater is a major component of the environment and it is not surprising to find, therefore, that one of the major challenges facing water management is and will be the need to minimise the environmental disruption society causes through its use of water.

On the one hand, the building of dams and reservoirs, especially large dams and reservoirs, has come under mounting criticism in recent years. Part of this criticism stems from the poor performance of the development programmes with which the construction of the dams was related especially, but not only, for irrigation. Much of the criticism, however, is also based on the failure to consider the environmental and social impact of the reservoir on the region in which it has been inserted and on the general absence of participation by the population affected in the process of design, construction and operation.

On the other hand, changes in the spatial distribution and structure of human activities, particularly increasing urbanisation, will continue to have serious effects on the water resource due both to the impact on patterns of streamflow and on water quality.

Equally significant, from the viewpoint of the role of the water resource in the environment, are changes in economic structure. Industrial growth and changes in industrial structure have been of particular importance, and the recent adjustments will only result in greater industrial growth in the decades ahead. Over the last 20 years, in nearly all economies, the manufacture of intermediate and capital goods has become as important as the production of food and other non-durable goods. The intermediate and capital goods industries and mining all demand large volumes of water in the production process and produce larger and more complex waste discharges.

Concern for the impact of economic development on the natural environment together with the increasing awareness of the close interrelationship between poverty, especially rural poverty, and environmental degradation has placed environmental management in the forefront of political discussion. The recent tendency towards rationalisation, decentralisation and privatisation of former public sector responsibilities in many countries has, however, brought about an unprecedented change in the institutional environment for water management as many new actors enter into the management and decision-making process. Management processes that were closed are now open, and general public discussion of water management decisions is increasingly common in many countries as the traditional centralised approach to decision-making is questioned.

An important aspect of the criticisms made of the traditional approach to decision-making is based on the lack of consideration given to the environmental consequences of water management decisions and the consequent damaging environmental effects of many decisions to construct works and assign water use. It is arguable that the over-centralisation of any activity is likely to lead to sub-optimum decisions and, especially, to a failure to consider their wider implications. A consequence is the ignoring of the environmental effects or impact of decisions. The more open and participatory the process of decision-making is, the more probable it is that all aspects of the decision will receive consideration. Obviously, this does not mean that perfection can be achieved, but only that better decisions will result from local decision-making.

The transfer of responsibilities from central government agencies to lower levels of government and to the private sector is producing a need for new institutional structures for water management. Centralisation of water management destroyed the traditions of local and user participation in management nearly everywhere. Now processes of decentralisation demand that the idea of user participation, of partnership among public institutions and the members of the private sector, be recreated through the adoption of institutional structures appropriate to the traditions and idiosyncrasies of local communities.

Privatisation leads, by its very nature, to an increase in the participation of non-government agents in water management. In itself, however, as has been emphasised, the transfer of responsibilities to the private sector will not be sufficient to create a new institutional system for water management. Such a system must be specifically created. The need for innovation is now widely recognised, but there has been only limited progress in the construction of systems of water management based on local institutions with wide social participation and where the environmental aspects of water management can be given their due weight.

Throughout the 20th century, the demands over the water resource have gradually intensified in most countries. This intensification of demands can be expected to continue as population continues to increase and economies continue to grow. Demand will not only continue to increase, but its nature will change as economic and social structures change. The changing and multiple roles demanded from the water resource will place tremendous pressure on the ability of managers to cope with the continually changing issues that must be confronted in water management. The earth's supply of freshwater is clearly compromised by these developments. There must, however, be no doubt that if we wish to successfully conserve and safeguard the precious nature of water as a resource, then we must learn to treat it as an economic commodity.

REFERENCES

Rogers, Peter (1992), *Comprehensive Water Resources Management. A Concept Paper*, Working Papers, Infrastructure and Urban development Department, Washington, D.C.: The World Bank.

Scudder, Thayer (1994), 'Recent experiences with river basin development in the tropics', *Natural Resources Forum*, **18** (2), 101–113.

Index

References in italic indicate tables or figures.